Thaddeus William Harris
(1795–1856)

Thaddeus William Harris (1795–1856). From: *Entomological Correspondence of Thaddeus William Harris, M.D.*, edited by Samuel H. Scudder (1869). Courtesy of the Ernst Mayr Library of the Museum of Comparative Zoology, Harvard University.

Thaddeus William Harris (1795–1856)

Nature, Science, and Society
in the Life
of an American Naturalist

Clark A. Elliott

Lehigh
University
Press

Bethlehem: Lehigh University Press

Associated University Presses
2010 Eastpark Boulevard
Cranbury, NJ 08512

The paper used in this publication meets the requirements of the American National Standard for Permanence of Paper for Printed Library Materials Z39.48-1984.

Library of Congress Cataloging-in-Publication Data

Elliott, Clark A.
    Thaddeus William Harris (1795–1856) : nature, science, and society in the life of an American naturalist / Clark A. Elliott.
        p.   cm.
    Includes bibliographical references and index.
    ISBN-13: 978-0-934223-91-1 (alk. paper)
    1. Harris, Thaddeus William, 1795–1856.   2. Entomologists—United States—Biography.   I. Title.
QL31.H3.E45   2008
595.7092—dc22
[B]   2007022896

To the Generations of Harvard Librarians Who Followed Harris
in the Enrichment of Learning and Advancement of Knowledge,
One Volume, One Bibliographic Record, One Research Response,
alas, One Committee, at a Time.

Contents

Preface

AMERICAN SCIENCE CAME OF AGE IN THE ANTEBELLUM PERIOD, ALthough full maturity would take at least until the end of the nineteenth century. Aspiring to a scientific career was a novelty in the early decades, but it got increasingly more routine as the century progressed. While significant work was done by the ever-growing American scientific community, few would argue that it rivaled the achievements of the older research establishment in Europe.[1] To be sure, it is "the *knowledge* that . . . [scientists] use and create that marks them off as distinctive practitioners and professionals,"[2] but in the absence of a supporting social structure the opportunity for sustained scientific investigation in a democratic society is limited. In antebellum America, sources of patronage were largely absent and remunerative work was a near-universal requirement.[3] Science is an inherently social activity, and it was in the difficult work of organizing and institutionalizing the practice of science that the antebellum period was engaged and which gave it its special character. Broadly speaking, the range of organizational activities encompassed educational preparation, communication channels and venues, employment opportunities relating to growing specialization in research, concrete economic and political support, and a sustaining social ideology that recognized the value of scientific research. At the outset of the antebellum period virtually all of these were largely absent in the United States.

While the persona of the expert who wields authority based on university training came into full flower only during the later Progressive Era,[4] academic institutions began to adapt their traditional liberal arts curriculum to the needs of science by about 1820 and the process "was essentially completed by 1850."[5] During the antebellum period, there were also developments aimed at meeting the special needs of science by establishing new institutions or separate schools of science within existing universities.[6] Doctoral programs arose after the Civil War, and by the early years of the twentieth century the vast majority of leading American scientists had earned the Ph.D.[7]

9

Learned societies played a crucial role in the promotion, coordination, and communication of scientific work, beginning with the eighteenth-century formation of the American Philosophical Society and the American Academy of Arts and Sciences. Establishment of the Academy of Natural Sciences of Philadelphia in 1812 is a milestone in the emergence of more specialized societies, as was the founding of the Lyceum of Natural History of New York in 1817 and the Boston Society of Natural History in 1830, successor to the then-inactive Linnaean Society of New England. In addition to specialization, scientific societies in the antebellum period also began to develop along national as well as local or regional lines. Formed in 1840, the scope of the Association of American Geologists expanded two years later with the addition of the term "and Naturalists" to its name. Before the decade ended, the group had become the American Association for the Advancement of Science.

In this time period, the scientific societies often became the focus of growing tensions between amateurs and the emerging professionals in science.[8] In the Academy of Natural Sciences, for example, the development of more professionalized approaches in natural history relegated the earlier explorer-naturalists to a subordinate position.[9] The Boston Society of Natural History, "founded to bring self-edification to its members but also to give focus to individual activities," continued, while competing venues for the professionals developed at Harvard University and in national organizations. In consequence, the Boston society evolved over time so that its amateur base emphasized public education rather than scientific research.[10] At mid-century, the American Association for the Advancement of Science became the central locale for the debate about amateurs and professionals, and over the question of who was to control the destinies of science in America. This contention was played out especially in regard to control over what got published from among the offerings at the Association's national meetings. The question came to a crisis in 1853 and the professionals prevailed.[11] In the antebellum period, the scientific paper became the primary means of publication of research results,[12] and the development of such publication outlets was an important part of the emerging infrastructure during this period. The Academy of Natural Sciences of Philadelphia, for example, began issuing its *Journal* in 1817, and through a process of peer review the academy was able to control access by the increasingly out-of-fashion field naturalists.[13] The *American Journal of Science* first appeared in 1818 and remained the most important publication throughout the antebellum period.[14]

In addition to their part in the education of scientists, the colleges and universities also played an important and expanding role in their employment. One historian has estimated that employment opportunities as professor of science grew by a factor of ten in the period from the late 1820s to 1860.[15] Teaching was still the salient mission of the colleges,[16] but Stanley Guralnick argues that "the research ethic had [become] . . . characteristic of the academic scientific community by mid-century."[17] Given the substantial role that academics played as contributors to the scientific literature,[18] Guralnick was undoubtedly right, although it does not answer the question of whether the colleges considered research as part of their *institutional* role. During the antebellum years other employment opportunities for scientists also emerged. Natural history, geological, and topographical surveys sponsored by government were an increasingly important presence. The first state geological survey was undertaken by North Carolina in 1823. The decade of the 1830s has been referred to as "an era of public surveys," and that of Massachusetts took the lead in 1830 after the North Carolina precedent.[19] In that time of developing values and rationale for scientific research, debate regarding government support raised questions about the relations of pure and applied science and whether the public should support an effort that did not have an economic benefit. Scientists generally wanted to pursue their study of geological or other questions, while economic interests (generally the lead motive for a survey) wanted to concentrate on the discovery and development of natural resources. These differing approaches sometimes resulted in public conflict.[20] In comparison to the academic and governmental sectors, commercial and industrial concerns played a small part in the employment of scientists in the antebellum period.[21]

The development of ideological support for science in American society was an important aspect of the history of that period, although its full fruition did not come until the post-Civil War decades. That science was more than a local activity came to the fore through awareness that Americans had a long way to go in order to achieve equality with Europe. Historian John Greene argues that the "colonial" period in American science had passed by 1820 though parity would come slowly over the next century.[22] The relationship was a complex one. While leaders of American science wanted to emulate Europeans, they hoped to avoid depending on them for approval or adopting European practice without appropriate adaptation to the American scene.[23] European attitudes to American science in the antebellum period could be critical and condescending, but native scientists and their supporters took up the challenge.[24] The rivalry with Europe was commonly used, for example, as an argument for government support of science.[25]

In certain respects, those trying to establish a niche for science in antebellum America had to fight on two fronts. While contending with the threat and the promise of European science on the one side, they also had to ward off interference from the general population in that era of Jacksonian democracy[26] and from other cultural forces, especially religion. In the antebellum period, natural theology—seeking evidence of the Divine in nature—emphasized the congruency of science and religion as ways of knowing God, and helped to fend off open conflict between the two. While many scientists were genuinely religious,[27] the appeal to natural theology by scientists during the antebellum period (especially in the period from about 1846 to 1860) indicates the relative dominance of science by religion during those years.[28] In the years after the war, natural theology became less important as a means of justifying scientific research,[29] and in time science tended to overcome religion as the leading cultural force.[30]

Science, as with all social organization, is a working together of institutions, ideology and cultural values, and actors. Structuring and defining the role of the actors was a major aspect of antebellum science in the United States and has been examined by historians as part of the general professionalization of work in society. This was a complicated historical process and did not happen all at once. At the level of ideology or self-belief on the part of the scientific community (as opposed to social or organizational structure), historian George Daniels identified several generalized stages of development toward professional status: (1) "*preemption*" involved the taking over by scientists of tasks formerly the prerogative of another group or the general population and was related to the increasing complexity of scientific knowledge that required special know-how in order to understand and interpret; (2) "*institutionalization*" entailed the development of identifiable ways of behaving as a mark of a true scientist, and was manifested especially in relation to the creation and operation of scientific societies; (3) in moving toward autonomy, "*legitimation*" required that scientists establish the worth of their activities, and in the antebellum period this involved an appeal to extrascientific values, such as the practical usefulness of their knowledge products, or the religious and moral outcome of studying science. It was later in the century before scientists were able to set aside such external justifications and to win sufficiency for the advance of physical knowledge as an acceptable end in itself.[31] Historian Elizabeth Keeney, for one, has argued that motivation for scientific work was the crucial difference separating the professional scientist from the amateur, at least in botany. The salient outcome for the professional was the advancement of knowledge, for the amateur personal "enrichment" (that is, self-improvement, religious experience, exercise, and the like).[32]

Defining and identifying members of the nineteenth-century scientific community has been a difficult task for historians and in itself illuminates the nature of the situation. Different studies have emphasized publication, membership and leadership in scientific societies, inclusion in standard reference works, and involvement in correspondence and data-collection networks. Depending on what criteria are used to define the scientist, the community can look quite different. It is clear that the population of individuals actively interested or involved in science in nineteenth-century America was larger than initial studies indicated.[33] Based on a much more select group of scientists born in the period 1776 to 1815,[34] the "typical" scientist was a college graduate, and may have held the M.D. in addition to the A.B. The life sciences were the most likely area in which a scientist would work in this era and together with the earth sciences demonstrate the importance of the natural environment in America as a focus of study. Though it is difficult to get a fine-grained picture of specialization at the level of subfields, it appears that, among the ranks of leading scientists, most researchers confined themselves to one general area of science; subfield specialization was particularly noticeable in biology, where work might be confined, for example, to ornithology or entomology.[35] Historian John Lankford argues that the decade of the 1850s was the period of transition in the emergence of a social role for the scientist and the development of national institutions of science, while specialization became important particularly in the decade of the 1860s and thereafter.[36]

Scientists found employment in a variety of occupations and positions. Many held multiple jobs during the course of their careers, often in situations unrelated to science. Nevertheless, science and science-related employment typified the scientist and, in comparison to other occupations, a professorship in science was particularly important, although only a little more than a third of the individual scientists ever held such an appointment.[37] Government at various levels also played a significant role as employer of scientists.[38] Most publishing scientists worked at some point in their lives in a science or science-related occupation and those working *only* in these categories were a majority of all scientists.[39] The historic trend from the colonial period to the twentieth century was a decrease in nonscience employment categories and the specific rise of the professorship in science. Though to a lesser extent than later, in the antebellum period the professor can be designated as the "typical" scientist of the era. Nevertheless, despite the obvious development of recognized educational patterns, field specialization, and employment networks—all of which would become mandatory for the professional scientist by the end of the century—in the antebellum period science in both its organizational and cultural dimensions still had close ties to the general society. [40] Prescribed avenues for preparation and success in science were not yet laid out, and while statistical patterns can now be uncovered, they would not necessarily have been all that obvious to the aspiring scientist at the time. In this context, how an individual dealt with contingencies in the pursuit of a life in science becomes a central role of antebellum biography.

This work concerns the life and career of one naturalist who faced those "contingencies" in an attempt to pursue scientific interests while meeting his economic and psychic needs. Occupational status is a central aspect of the definition of a professionalizing scientist, and the employment of biologists differed somewhat from that of the general population of scientists discussed above. The percentages for nonscience, and science and science-related, occupations were virtually the same as the general population. However, medicine was the leading occupation among naturalists, surpassing professorship in science, which was the largest occupational category for the scientific population overall.[41] The close association of medicine with biological study is hardly surprising. Furthermore, historian Stanley Guralnick reports that natural history was the last of the scientific fields to be included in the antebellum college curriculum, and the figures given here no doubt reflect the relative lack of academic opportunities for naturalists during that time period.[42]

The cohort of American zoologists and botanists active during Thaddeus William Harris's lifetime that were used above as a basis for the biologists' employment figures included 184 individuals, a fair representation of the more active researchers during those years. Their commitment to biological inquiry varied, of course, as did their life stories. Harris's chief publication was the *Report on the Insects of Massachusetts, Injurious to Vegetation* (1841),[43] published under the auspices of the Massachusetts Commissioners on the Zoological and Botanical Survey of the State. A brief look at the individual careers of the other six commissioners gives a glimpse of the avenues to natural history study during Harris's lifetime.[44]

Ebenezer Emmons (1799–1863), known chiefly as a geologist, was assigned the quadrupeds for the Massachusetts state survey, referred to as "the first modern account of a state's mammalian fauna."[45]

Born in Massachusetts, he graduated from Williams College and from Rensselaer Polytechnic Institute and also studied medicine. He was a geologist with the New York Natural History Survey (1836–42) and later served as custodian of New York state collections and participated in the state agricultural survey. In 1851, Emmons became North Carolina state geologist. In addition to these appointments he continued to practice medicine, and during most of his adult years he was also professor of natural history (later, geology and mineralogy) at Williams College as well as teaching chemistry for shorter periods in the Albany Medical College. Some of his geological interpretations were controversial, and the wide opposition that resulted prevented his full and positive involvement with the contemporary scientific community.

T. W. Harris's friend, Augustus Addison Gould (1805–66), took charge of the state's invertebrates (other than insects) and, like Harris, established his reputation as a naturalist with the work he did on the state survey. Gould was born in New Hampshire and graduated from Harvard College and the Harvard Medical School (as did Harris). He was one of the leading physicians of Boston and pursued his conchological studies in moments not devoted to his medical practice. His survey report was the first regional study of mollusks done in the United States and covered terrestrial, freshwater, and marine specimens; unlike Harris, Gould did not pursue the economic or applied aspects of his subject. His reputation was such that he was called on to prepare reports on conchological specimens collected by United States government exploring parties, including his important work on the Wilkes Pacific exploring expedition collection. Gould included the study of geographical distribution of species in his work on mollusks. At a time when Harvard was without a professor of natural history, Gould had responsibility for teaching the subject in the College for two years before Harris took over the assignment. Gould, with Louis Agassiz, published *Principles of Zoology* (1848), a textbook for use in schools and colleges. A posthumous second edition of his state report was issued in 1879.

The birds of the state were assigned to William Bourn Oliver Peabody (1799–1847), a seemingly unlikely choice. Born in New Hampshire, he graduated from Harvard College the year after Harris and, following a year of teaching, began preparation for the ministry. He spent his entire career as pastor of a Unitarian church in Springfield, Massachusetts. Peabody reportedly wrote extensively for periodical publications and also published sermons and poems. His writings included contributions to Jared Sparks' *Library of American Biography*, among which was a life of ornithologist Alexander Wilson, and he knew John James Audubon. His brother wrote of him that "he found a relief from his severer toil in the contemplation and study of nature," and was interested in all branches of science, although he studied those that his circumstances made most feasible.[46] Of his report for the state survey, his modern biographer notes that it was essentially an addition to Thomas Nuttall's *Manual of the Ornithology of the United States and Canada* and followed the expectations of the survey by stressing the economic aspects of the state's birds.[47] While studying the habits of birds was a special interest for Peabody, it is apparent that the subject was one among a number of intellectual interests, a situation that distinguishes him from the other survey commissioners (other than Emmons), for whom the assigned department of natural history constituted a committed specialization.

David Humphreys Storer (1804–91), another friend of Harris, worked on the state's fishes and reptiles and used the opportunity that the state survey offered to lay the foundation for his life's work in ichthyology. Storer was born in Maine, still a part of Massachusetts. After graduating from Bowdoin College, he went on to earn a degree from Harvard Medical School. In his lifelong Boston medical practice, he specialized in obstetrics and helped found a private medical school in Boston. He later held a professorship in obstetrics and medical jurisprudence at the Harvard Medical School, where he also served as dean. Prior to the time he was appointed to the state survey commission, Storer's chief zoological interest had been in mollusks (assigned to Gould) so that his turn to ichthyology was an outcome of his work for the state. After he submitted his report, he continued for many years to work on a revision and expansion of the part relating to fishes. Published serially in the *Memoirs of the American Academy of Arts and Sciences,* it finally appeared as a separate volume in 1867. Not a field worker or collector of fish specimens, he was assisted by fishermen and others involved in commercial fishing and was concerned with the economics of fishing in addition to descriptions of individual species. Storer also published extensively on medical topics.

One of the two botanists on the survey, Chester Dewey (1784–1867) reported on herbaceous flowering plants. Born in Massachusetts, he graduated from Williams College and then studied for the ministry. His career was devoted to teaching at both the college and secondary school level. He was professor of mathematics and natural philosophy at Williams College, where his main teaching responsibilities included botany and chemistry, as well as mathematics, geology, and astronomy. Ebenezer Emmons was one of his students at

Williams. Subsequently serving as principal of a gymnasium in Pittsfield, Massachusetts, at the time of the state survey Dewey was a high school principal in Rochester, New York. At various points in his career he was also a medical school lecturer in chemistry and medical botany. He ended his career as professor of chemistry and natural sciences at the University of Rochester. Though Dewey's teaching responsibilities were wide-ranging, his published contributions were chiefly botanical, especially on the Carices (sedges), which appeared chiefly in the *American Journal of Science* over a period of some forty years.

George Barrell Emerson (1797–1881), as president of the Boston Society of Natural History, headed the Commissioners on the Zoological and Botanical Survey of the State, and took charge of trees and shrubs. The same year that the commission was established, Emerson also took part in initiating the state board of education. Born in Maine (Massachusetts), he graduated from Harvard College two years after Harris. He followed a career as a teacher and educator. After serving as the first principal of the public Boston English Classical School, Emerson established a private school for girls. His report on trees and shrubs became the focus of his scientific activities and had a two-volume third edition more than three decades after its first issuance. As with Harris's report on insects, the work was praised for its balance of sound science and popular appeal. Emerson also published widely on educational topics

These summary portraits of Harris's coinvestigators on the Massachusetts survey show the important role that medicine played in antebellum natural history (Harris was himself originally a physician). Furthermore, it appears that nonscience employment was relatively more important among the survey scientists than for Harris's larger naturalist cohort (Harris was a university librarian during most of his adult years). On the other hand, professorships in science seem to have been less germane to the work of the survey than in the naturalist community in general. Emmons was professor at Williams College at the time of the survey, and Dewey had been (and would again be) a professor but did not hold such an appointment at the time of the survey. Unlike Harris (as will become evident), it does not appear that others on the survey—namely, Gould, Peabody, Storer, Emerson—aspired to a professorship in science and were reconciled to carrying out their biological research while practicing medicine, preaching, or teaching. Most of the surveyors either were, or became, remarkably specialized in their scientific work, although Emmons was really a geologist rather than a mammalogist, and Peabody was likely an amateur in science in the strict sense of the term. All, of course, were by definition engaged in government-sponsored science if only for this restricted time period; government science was a significant aspect of Emmons's career as it was, to a lesser extent, for Gould.

It is clear that Thaddeus William Harris shared much with the local and national scientific community of which he was a part for more than three decades. As indicated in the discussion above, his was a time of change, of optimism, of growth and development in American science. It was also a time of limited opportunity. Harris's story is interesting, significant, and, in some aspects illustrative, demonstrating as it does the interplay of limited social support with "personal drive, discipline, and ability."[48] The value of biography is twofold, in that individuals are inevitably both representative and unique in history. As historian of science Charles Rosenberg states the case, "although every life is idiosyncratic, no life is random."[49] This biography of physician, librarian, and entomologist Thaddeus William Harris is written in that spirit. While all aspects of his life are explored, emphasis is on his aspirations and work as a naturalist.

A broadly thematic approach facilitates the narration of Harris's life while also exploring, by example, some conditions and problem situations that many other American scientists would have faced in the antebellum period. The first chapter is a conventional examination of Harris's birth, family, and education, but also pays particular attention to the origins of his lifetime interest in science, and entomology in particular. Chapter 2 deals with Harris's employment and especially his attempts to find a position that would allow him to pursue his entomological researches. All naturalists in this era faced the problem of lack of collections and books to support their work and the ways in which Harris attempted to deal with this common problem is explored, including his contributions to research resources that were shared by the community of scientists. The next chapter (3) considers Harris's scientific work more specifically, within successive periods of his life. Harris is generally considered the "founder of applied entomology" in the United States, and his historical reputation links to his 1841 report for the Massachusetts Zoological and Botanical Survey of the State. Going beyond that work and its historical significance, however, and examining Harris's life through his extensive correspondence reveals a more complex picture than his traditional reputation would suggest. In this reading, he becomes less an agricultural entomologist and more a well-rounded student of insects in all the manifestations of the field. The chapter also reviews Harris's interest in botany and in some nonscience topics, and ends with a review of his tenure as Harvard University librarian, including methodological connections between his entomological and library work.

Chapter 4 examines Harris's scientific activities from a social (community) perspective and in doing so incidentally touches upon concerns that all naturalists of the era would encounter in one way or another. Among the questions that arise are Harris's views on popularization of science which, in his case, also touches on the relation between doing science for its own sake and the relevance of entomology to agricultural interests. Americanist historians generally recognize that scientists of the antebellum era did not conceive of a divide between scientific study and the usefulness of scientific knowledge,[50] although in practice the picture was still a complex one. Some of these issues are considered in this chapter while exploring Harris's development of an approach to journal article publication. Here he tried to differentiate between scientific and agricultural ends and, in that context, to differentiate between pure and applied science as it related to his vision of himself as a scientist.

Harris had well-formed views on standards and procedures for carrying out work in natural history, especially the naming of newly discovered species, and a discussion of these interests is a contribution to our understanding of biological practice in the period, including the patriotic-nationalistic outlook that motivated Harris and other American scientists of the time. The chapter also explores Harris's relations with a whole range of entomologists, including fellow investigators and individuals who were essentially suppliers of specimens and information on insects. While this section shows Harris at work it also demonstrates concretely how natural history networks operated in the first half of the nineteenth century, both domestically and internationally. The chapter ends with a narration of Harris's involvement with formal scientific organizations, especially the Boston Society of Natural History.

Harris's political and religious outlook is explored in chapter 5. The interplay of ideology, as derived and adapted from the general social and cultural milieu, and its relations to scientific practice, is something that historians need to consider, and one purpose of this chapter is to show how complex it can be even in the life of a single individual. While Harris was opinionated on political matters, in fact he was little involved in that sphere. His opinions, however, suggest how his interest in agriculture, which he considered the foundation for a stable society, underpinned aspects of his entomological program. Consideration of his religious views widens into a discussion of his attitudes to nature and how the objects of his scientific study were on occasion approached through the perspective of natural theology. At a less spiritual level, the chapter considers ways in which Harris integrated nature study into his distracted life. Never a mere collector, his great interest in the study of life histories of insects and the relations of animals and plants that such an approach entailed, gave him an ongoing interest in nature not only as locale for scientific study but as scene of life. Finally, chapter 6 deals with Harris's death and his long-term reputation. It especially addresses the question of how his personal conception of his career in science relates to what history subsequently made of him and his work. In this regard, it explores particularly the relations of his work in basic taxonomics and his work as an applied entomologist. In the fluidity of scientific roles in the antebellum period, this general question of individual vision and social uses of a scientific reputation should be broadly relevant.

Acknowledgments

HOPE THE FOREGOING WILL SUGGEST SOMETHING OF THE INTRINSIC
value that a biography of Thaddeus William Harris holds and that
some of the more general questions will interest readers as they
work their way through the details of his life. Getting to this point in
an academic study requires some autobiographical input of my own.
Much of my adult professional life has been devoted to the amassing
of biographical data on American scientists of the nineteenth century both for individual sketches, as in my *Biographical Dictionary of
American Science,* and several quantitative works.[1] When I decided to
turn my attention to a particular life, my interests were drawn to a
research scientist who seemed to be close to the workings of the natural world (Harris was noted for his interest in the life histories of
insects), and I was working in the Harvard University Library (where
Harris had once presided). Given the reasonably substantial collection of correspondence and other papers that Harris left behind,
the decision was an easy one. Carrying it out has been a project of
many years, with successive starts and stops, as other projects have
intervened or as I just got stymied on how to do Harris's life justice
while making it relevant to those interested in the early history of
science in the United States.

Over the years I have benefited from the interest and kindness of
many professionals and by the helpfulness and encouragement of
friends. Barbara Wiseman and the late Edward (Ned) Pearce at the
Boston Museum of Science library took the lead in time by introducing me to the Harris Papers then under their care. On successive
occasions they accommodated my interests and facilitated my use of
the collection and it was then, in the early 1970s, that the bulk of my
research was done in this primary cache of Harris Papers. Since the
collection was transferred to the Museum of Comparative Zoology
library at Harvard University, Ann Blum, Robert Young, and Dana
Fisher have been my lifeline not only to that collection but to a number of other Harris letters and documents in their curatorial care,
so readily accessible through the library's finding aids and indexes.
My former colleagues at the Harvard University Archives have been

21

invariably helpful, notably Patrice Donoghue, Robin McElheny, and Brian Sullivan. A particularly heartfelt thanks goes to Brian, now an independent scholar, not only for his professional assistance in accessing the treasures of the archives, but for his perennially enthusiastic support, and his willingness to share his own work on John Langdon Sibley, Harris's assistant and successor as head of the Harvard library. The staff at the University's Houghton Library efficiently delivered and facilitated use of several manuscript collections relating to Harris, for which I am both grateful and confirmed in my admiration for the value of that great historical and literary institution. I am likewise indebted to Judy Warnement and staff for assistance in using the manuscript resources at the Harvard Botany Libraries. Over the years I have had the advantage, and always the pleasure, of using the Boston Public Library and the libraries of the Massachusetts Historical Society and the Massachusetts Horticultural Society. Aurore Eaton at the Cambridge Historical Society and Edith Clifford with the Milton Historical Society were both interested in my project and willing to help in accessing resources or tracking down information. In the later stages of the research, several libraries offered exemplary service in providing photocopies of vital documents, namely the Academy of Natural Sciences of Philadelphia (Carol Spawn, curator of manuscripts and archives), the American Philosophical Society (curator Robert Cox), and the Louisiana State University Libraries Special Collections department. Carolyn Kirdahy, current librarian at the Boston Museum of Science completed the circle by facilitating my access to the records of its parent organization, the Boston Society of Natural History.

The late Frank Morton Carpenter, professor of zoology at Harvard University, deserves a special public acknowledgment of his typically generous role in the early stages of this project. I made his acquaintance when I audited his introductory entomology course and he became interested in and encouraged my undertaking. His support was also expressed in a very material way when he made me the gift of his personal copies of Harris's *Treatise on Some of the Insects . . . Injurious to Vegetation* (1842 and 1862 editions), and *Entomological Correspondence* (1869). They are among my most prized possessions. Historian Sally Gregory Kohlstedt, then on the faculty of Simmons College in Boston, was from the first a resource for information on the whereabouts of Harris manuscripts and a valued friend who encouraged the undertaking. Margaret Rossiter and Joy Harvey are among other historian friends who have given much welcomed support over the years. I am indebted to Conner Sorensen for both his personal interest and his 1995 *Brethren of the Net,* a work that thoroughly and imaginatively delineates the history of American ento-mology in the nineteenth century. It was vital to my research and makes me glad I waited to finish the biography. Hunter Dupree's life of Asa Gray was an important source of information as well an instructive example of scientific biography at its best, and Hunter's friendship and interest in the project were at all times a source of reassurance that it was worth pursuing. Fred Burchsted in the Wide-ner Library at Harvard University probably knows more than anyone about the naturalists of New England, and he has generously shared that knowledge along with his command of the general literature on the history of science. Anthony Sammarco, at the House of the Suf-folk Resolves in Milton, Massachusetts, where Harris once lived, was helpful with both Dorchester and Milton history and his interest and knowledge about Harris.

As with all ordinary things in our life together, I hardly know how to thank my wife Priscilla. Harris has been in my life for so long that he has become a part of it, looking out from his frame on my desk and reminding me of a commitment made. Priscilla never urged quitting or getting on with the writing but patiently and often si-lently lent endorsement, because she knew the project was impor-tant and meaningful to me. It has meant much to me in this instance as in so many others to be assured that I had her understanding and support, which I guess is what forty-plus years of marriage is all about. Ours sons Andrew and Glenn have likewise kept with me on the project over many years. For all these friends and family, and others I may have overlooked, thank your for the up-lifting.

Thaddeus William Harris
(1795–1856)

1

Origins

BACKGROUND

THADDEUS WILLIAM HARRIS, PHYSICIAN, LIBRARIAN, AND ENTOMOLOGIST, was born on November 12, 1795, in Dorchester, Massachusetts, the first child of Thaddeus Mason Harris and Mary (Dix) Harris. His father was a Unitarian minister. Dorchester, just south of Boston (and now a part of the city) had about 305 houses in 1800, with a population of 2,347. In the early 1820s it still was described as an agricultural area well settled with both farms and country homes. A mid-century historical account commented that "The location of Dorchester is picturesque, and even elegant," the "surface of the ground . . . uneven enough to give that agreeable variety of hill and dale so charming to a poet's eye."[1] Harris himself says little about his childhood, but according to his daughter he always longed for "a home in the middle of a ten-acre lot," an ideal that went back to his early life at Mt. Ida in Dorchester, where the house stood at the center of seven and a half acres of land.[2]

Though the personalities of Harris's parents differed considerably, they both had an important influence on him. The character of his father, Thaddeus Mason Harris (1768–1842), was emotional, moving between depression and elation yet he was described as "invariably modest, gentle, compassionate."[3] He presided over the First Church in Dorchester from 1793 to 1836 while pursuing a variety of scholarly projects. For two years before going to Dorchester, he was librarian of Harvard College (thus preceding his son in that position), and after retirement from the ministry he was librarian at the Massachusetts Historical Society (1837–42).[4] He published fifty-eight sermons and seventeen other works,[5] including *The Minor Encyclopaedia* in four volumes. He was a member of the Masons, published on that organization, and was much maligned in the press after the abduction and disappearance of William Morgan in New York state in 1826, an event that gave rise to anti-Masonry as a political movement.[6] In an effort to recover his health, Rev. Harris took a four-month trip to Ohio in the spring of 1803, an account of which he published as *The Journal of a Tour into the Territory Northwest of the Alleghany Mountains . . . With a Geographical and Historical Account of the State of Ohio* (Boston, 1805).[7] He also published *The Natural History of the Bible,* which first appeared in 1793 with a more extensive version in 1820, subtitled *A Description of All the Quadrupeds, Birds, Fishes, Reptiles, and Insects, Trees, Plants, Flowers, Gums and Precious Stones Mentioned in the Sacred Scriptures, Collected from the Best Authorities* (Boston).[8]

From all evidence, Harris and his father continued to have a close relationship throughout their lives. They shared interests as well as a natural connection and the "loss of a beloved & venerated father" in 1842 was an event of considerable importance in Harris's adult life.[9] While both gained distinction, although less from any inherent genius than through shared qualities of "untiring industry . . . thoroughness, precision, and accuracy in their literary work," Harris's own son characterized his father and grandfather as widely different in character. Harris's personal qualities are, in fact, attributed to those of his mother.[10]

Mary Dix (1776–1852) was the only daughter of Dr. Elijah and Dorothy (Lynde) Dix, whom the future Rev. Harris met while teaching temporarily in Worcester, Massachusetts.[11] Elijah Dix (1747–1809) was both a physician and druggist and an ambitious man who engaged in a number of civic and business enterprises including an interest as founder of the Maine towns of Dixfield and Dixmont. He died in the latter place under what apparently were somewhat mysterious circumstances.[12] Mary Dix Harris is described as imperious and controlled, one whom "Nature had endowed . . . with a commanding person, unusual intelligence, and great force of character," and whose eyes "may have seemed rather to penetrate the thoughts of others than to seek to win their affections," a persona set in "striking contrast to the meek and yielding spirit of the man she had chosen to wed."[13] Harris's own son, in fact, presents his father and grandmother as much alike and close because of this, Harris taking from her his "force of character, with a certain degree of sternness" which the son thought may have become more apparent in later years.[14]

The fact that Harris's father was a Harvard graduate, settled minister, and scholar gives good indication of the social status of Harris's upbringing, but his financial status is somewhat more problematic. Thaddeus Mason Harris's own father had died at an early age and left his family with little means, and it seems that Rev. Harris always was troubled by financial concerns. Although his father-in-law had risen to a position of wealth, it appears that in his lifetime he was not always forthcoming with assistance. Reverend Harris's biographer relates the painful circumstances of his apparent hopes for financial assistance from his father-in-law, but when, early in his marriage, he undertook to construct "a large and expensive house" no such help was forthcoming, with the result that Rev. Harris suffered a considerable burden and embarrassment.[15] The house was built in 1793 and known as Mount Ida. It was here that Thaddeus William grew up.[16]

Harris was the first born of eight children. Two sisters, Sarah Duncan (1811–33) and Rebecca Mason (1813–17) pre-deceased him, as did a brother, Elijah Dix Harris (1798–1841), who was a merchant in Boston and St. Louis. When the memoir of Thaddeus Mason Harris was written in 1854, four of his children were still living in addition to Thaddeus William, namely Mary Dorothy (1797–1890); Clarendon (1800–84), resident for at least part of his life in Worcester, Massachusetts; John Alexander (1804–1882), a resident of Dorchester who died in Paxton, Massachusetts; and James Winthrop (1806–79), who resided in Cambridge and Dixmont, Maine, but only Thaddeus William was recognized for his accomplishments as Harvard librarian and as "one of the most distinguished entomologists of our country."[17] One of his brothers did at least occasional bibliographic and clerical work, in 1842 preparing a copy of an index to the *North American Review*,[18] and was engaged about 1848 in pasting and pressing what was probably the Harvard library catalogue.[19] This most certainly was brother James Winthrop, who also assisted Harris in preparations for moving the library to the new Gore Hall in 1841.[20] In 1847, brother Clarendon, in Worcester, was engaged in the settlement of their aunt's estate and was handling investments for Harris,[21] so it appears that he was a lawyer or businessman. In the mid-1840s, a sister also was engaged in clerical work with Harris as intermediary, working on the letterbooks of Jared Sparks. This would have been Mary Dorothy, who did not marry.[22]

EDUCATION

Little is known of Harris's precollege education. His Harvard class record book indicates that he prepared for college with David Brigham, Harvard A.B. 1810.[23] Harris's son reported that he prepared for college in Dedham, Massachusetts, and with the "kindly" Rev. Zedekiah Sanger at Bridgewater.[24] Sanger, who was minister of the

First Congregational church in Bridgewater, Massachusetts, had a particular interest in mathematics and science, and was known for his teaching activities in conjunction with his ministry. He was an original member of the American Academy of Arts and Sciences.[25] Harris later referred to Levi Leonard, who assisted him with his entomological work, as having been a "schoolfellow,"[26] and Leonard was born in Bridgewater so it is possible that the two were schoolmates before they both went to Harvard.

Harris entered Harvard in 1811. His roommate for his final three years was Charles Briggs, later a minister and secretary of the American Unitarian Association.[27] John Kirkland, who became university president the year before Harris entered, was epitomized as a Unitarian in religion and Federalist in politics, and as president "one of the most remarkable . . . that Harvard has ever had, and the best beloved."[28] During Kirkland's years, 1810-28, he greatly increased the number of professorships, and expanded the physical facilities of the college, including the construction of Holworthy Hall (1812) and University Hall (1815) during the years in which Harris was a student. The former was a residence hall (where Harris lived during his senior year).[29]

The curriculum during the time when Harris was a student continued in the vein it had followed for decades, virtually prescribed for all students. The primary means of instruction was through recitations, and as Bernard Bailyn has observed, the course of study represented "less subjects to be explored than books to be conquered."[30] An extract from the college laws gives an overview of the range of studies pursued, where annual public examination of students included "Latin and Greek Classics, Hebrew, Elements of English Grammar, Geography, Chronology, Antiquities, Ancient and Modern History, General Grammar, Logick, Rhetoric and Belles Lettres, and English Composition, Arithmetic, Algebra, Mensuration, plane Geometry and Trigonometry, Conic Sections, Surveying, Mensuration of Heights and Distances, Navigation, Dialling, Projections of the Sphere, Spheric Geometry and Trigonometry, with their application to Astronomical Problems, Fluxions, Natural Philosophy, the Elements of Chemistry, Astronomy, Metaphysicks, Moral and Political Philosophy, Theology, and such other studies pursued under the direction of the University, as may at any time be pointed out." The optional or elective studies were limited to natural history, French, and additional study in mathematics.[31] If the plea of Harris's contemporary, Stephen Salisbury (Harvard A.B. 1817), to his parents is a valid judgment, the course of studies and their arrangement was such that it kept the students very busy writing, translating, and calculating, as well as attending prayers and dealing with other devices of regimentation.[32] There is no record that Harris engaged in any formal extracurricular activities such as membership in student societies or groups. Among the books he charged from the university library was Richard Dean, *An Essay on the Future Life of Brute Creatures* (1768), in December 1812, and during his senior year he charged Spencer's *Faerie Queene* on several occasions, as well as William Paley's *Natural Theology* in June 1815, and "Winslow's Anatomy" in April, the latter during the period when medical lectures were delivered to seniors.[33]

Interestingly, during Harris's years at Harvard, in 1813, both his father and future father-in-law, Amos Holbrook, received honorary degrees from the university, the former an S.T.D and the latter an M.D. As part of his own performance, in November 1814 Harris participated (with three other students) in a so-called conference "on the influence of early impressions, speculative sentiments, popular opinion and professional occupation as forming the character."[34] Undoubtedly a highlight of Harris's years as an undergraduate was the end of the war with Great Britain, which was the occasion for elation and celebration at the College. In March 1815, one of his fellow students wrote to his family that on the news of the treaty "the air reechoed with frequent Huzzahs, the Colleges [residences] were decorated with flags of every Nation. Sixty flags were flying from the tops of the colleges." A dinner was held with the president and other dignitaries and, in the evening, an "illumination . . . of taste & elegance, [of] such brilliancy and splendor as far surpassed the illumination in Boston," with messages formed by candles in the windows of the buildings, Harris's Holworthy Hall showing "Union & Universal Peace."[35] Harris received the A.B. degree from Harvard at commencement August 30, 1815, and for the occasion was assigned a conference part (with his roommate Charles Briggs and another student) on the topic of "pastoral, epic and dramatic poetry."[36]

In the years immediately after leaving college he actively developed his interest in natural history and entomology. From March to November 1816 and March to December 1817, he taught school in Dorchester. He apparently attended medical lectures in Boston during the winter of 1816–17, and published records confirm that he attended the three-month long course of lectures in 1817–18, 1818–19, and 1819–20.[37] According to the published regulations of the medical school, in order to appear for the degree examination, which covered anatomy, physiology, chemistry, materia medica, pharmacy, midwifery, surgery, and the theory and practice of medicine, students were required to attend two courses of lectures. The degree also required the submission and public defense of a dissertation and stipulated three years of study with a "regular practicioner of medicine."[38] Harris studied with Dr. Amos Holbrook of Milton, a family friend and accomplished surgeon,[39] and received the M.D. degree from Harvard on March 2, 1820.[40]

ORIGINS OF AN INTEREST IN SCIENCE

Interest in science and related activities appear at various points in Harris's family background, perhaps most notably in his father's work on the natural history of the Bible. Work on the project spanned Harris's development to early adulthood, and his nineteenth-century biographer speculated that it "must have involved, on many . . . points, enough of minute investigation to enlist the whole family in the work."[41] His upbringing offered other inducements as well. Dorchester as a community was well-known for its horticultural interests, and Harris's father was among those devoted to growing fruits and flowers.[42] His father also was an eager joiner of "every association that was meant to promote the objects of natural science, or any good learning, or social improvement, or religion, or charity,"[43] and in 1829 both father and son became original members of the Massachusetts Horticultural Society.[44] On occasion, the particular interests of the two men coincided even more closely, as with Rev. Harris's 1821 article "On a Mode of Destroying Insects," which appeared in the *Massachusetts Agricultural Repository and Journal* (6: 392–93). While Harris's own commitment to natural history, and entomological studies in particular, developed well beyond his father's somewhat incidental involvement, Rev. Harris continued his interest and support. About 1837, at the time that Harris was appointed to prepare his report on insects for the commonwealth, Rev. Harris wrote a "dear son" letter offering his aid and advice, especially on means of controlling the damage caused by insects, with the suggestion and hope that "some observations which I have myself made, & some notices which I have met with, may be acceptable," proposing "from time to time, to make a statement of any facts or incidents that to arise [*sic*] to my recollection."[45]

Interest in entomology in Harris's family was not limited to his father. White mulberry trees grew at his parents' home and for many years his mother raised silk worms from which she supplied her own sewing thread. In 1828, Harris promised to send his entomologist friend Nicholas M. Hentz some of his mother's silk moth eggs.[46]

Though nowhere in Harris's papers is mention made of his relations with his maternal grandfather, it is entirely possible that a significant influence also came from that direction, and his closeness with his mother is one of the reasons to suppose it. Elijah Dix, as indicated above, was a physician and also a druggist. In the 1780s, Dr. Dix traveled to England and while there he made connections with the chemical trade and returned not only with chemical substances but books and apparatus as well. In 1795, the year of Harris's birth, Dr. Dix built a house in Boston, moved there, and engaged in the wholesale drug business.[47] With his interests and proximity to the young grandson, it is too much to suppose that he did not have an effect both in exciting young Harris's scientific interests and probably also his later pursuit of medicine as well. To the degree to which this is true, it had to have been an early influence, since grandfather Dix died in 1809, when Harris was thirteen years old.

Harris's active commitment to science began in 1823 with his first publication and his confession to Thomas Say of "an ardent love of Natural Science."[48] The exact chronology of the origin and early development of Harris's scientific studies, however, is somewhat uncertain because he says little that predates his college years and, surprisingly, never acknowledges the family influence outlined above. In 1852, he did say that since boyhood he had been engaged in the study of insect larvae and their transformations but does not mention an exact age,[49] and his early biographer, T. W. Higginson, doubted that he had any well-developed interest in natural history at the time he graduated from college.[50]

Harris attributes his particular scientific interests to William Dandridge Peck, the "esteemed friend who first developed my taste for entomology, and stimulated me to cultivate it."[51] Peck was Massachusetts Professor of Natural History at Harvard and curator of the Botanical Garden, and it has traditionally been assumed that Harris attended Peck's course on natural history when he was in the college.[52] Unlike the bulk of the curriculum at the time, however, the botanical and zoological lectures were optional, open to juniors and seniors at the request of their parents, and with the payment of an extra fee,[53] and it appears that Harris never was officially a student of Peck.[54] Whether or not Harris took the course in natural history, the influence of Peck on his interests apparently was exerted after he left the college in 1815.[55] He was cultivating a prepared mind, however, because there is indication of a serious interest in natural history as an undergraduate. For example, during Harris's final term in college he charged out the first volume of Buffon's *Natural History* from the library.[56] Also, while in Cambridge, either as a student or shortly thereafter, he studied botany with his classmate and friend

Levi W. Leonard (1790–1864),[57] a fact supporting both the likelihood that his interest in natural history arose independently of Peck and his later admission that his initial natural history interests were botanical rather than zoological.[58]

The exact origins of Harris's interest in medicine is uncertain, although a possible influence from his grandfather Dix was suggested above. Also, Dr. Amos Holbrook, who was a long-time friend of Harris's father, "regarded [Harris] from a youth as an adopted son" and became his medical teacher (and later his father-in-law).[59] Almost certainly, Harris was first interested in science and then in medicine. However, his son's 1882 memoir has a somewhat different interpretation of the direction of influence, writing, "During the preparation for his medical life he had paid some attention to the study of botany, and as early as 1819 we find him carrying on an animated discussion by letter with Dr. Dow, of Dover, New Hampshire, on botanical and entomological questions in their relations with Materia Medica. This is the first recorded evidence of a decided or pronounced taste in the direction which his studies of later years assumed."[60] The letters exchanged with Dr. Dow have not been seen, but the allusion to an "animated discussion" suggests not a beginner but a student already committed to his subject. More importantly, even at this stage of his career, it becomes apparent that Harris was not one to view his scientific interests in isolation but in relation to other concerns, and it is that interconnectedness, between plants and insects, between plants, insects, and medicine, and between plants, insects, and agriculture, that lay at the foundation of his scientific interests.

The difficult general question of origins of interest and Harris's initial motivation for the study of natural history aside, there is no doubt that it was William Dandridge Peck who gave early encouragement to him as a naturalist and most likely directed him toward a particular type of interest in entomological studies. As noted above, Harris himself was aware of that mentoring role, in later years recalling Peck as his "friend & patron."[61] Peck was the first American-born naturalist to engage in the scientific study of insects and, equally important, the first American economic entomologist.[62] The early historian of American entomology, John G. Morris, said of his output, it was "published in agricultural journals which had not a wide circulation out of New England, and difficult, perhaps impossible, to be procured at the present time [1846]."[63] His contributions to descriptive entomology were few, while he concentrated instead on "the economic aspects of the subject" with attention to insects that cause damage to vegetation.[64] Peck also contributed to ichthyology and botany.

Peck was not particularly notable as a classroom lecturer,[65] where he featured the Linnaean classificatory system.[66] Still, he did influence a group of students, in addition to Harris, who later pursued natural history. Peck's professorship and the related botanic garden were established with the sponsorship of the Massachusetts Society for Promoting Agriculture, and Peck was the candidate of the subscribers.[67] The official beginnings of natural history teaching and research at Harvard, therefore, reflected an agricultural interest, and Harris later pointed out in his correspondence that the professorship had "required by the statutes instruction particularly in this branch [entomology], on account of its close connection with botany."[68]

The Massachusetts Society for Promoting Agriculture played an important role in encouraging Harris's early entomological career. His first paper, on the natural history of the saltmarsh caterpillar, was published in 1823 in the Society's *Massachusetts Agricultural Repository and Journal,* and it won a prize from the society, an event that Harris credited to the intervention of John Lowell (1769–1840), until 1822 a member of the Harvard Corporation and a member of its Board of Overseers for several years thereafter.[69] Lowell was one of the most active members of the Massachusetts Society for Promoting Agriculture, serving as a trustee, secretary (and editor of the *Repository and Journal*), and as president from 1823–28.[70] It was Lowell who orchestrated the uncomfortable event that was Harris's transition from a protégé of Peck to a naturalist who, having learned from and having been encouraged by Peck, and who in important ways having taken up his entomological program, nonetheless, came to realize that the future required a naturalist of a somewhat different character than Peck had been.

Peck left to Harris the task of preparing his manuscripts for publication after his death,[71] which occurred in 1822. Harris's father also became involved in the task and Lowell apparently played a crucial role in the actual decision regarding publication. By about 1824, it had become apparent to Harris and his father that publication of the papers would neither aid Peck's reputation nor financially benefit his widow.[72] Lowell had to be informed of their conclusion, and Harris's letter is a revealing portrait and evaluation of Peck by his best known student. Although the name of the addressee of the following is not given, referring instead to "Respected & dear Sir," it almost certainly is addressed to Lowell. Harris wrote,

Not having it in my power to see you in person, I take the liberty of addressing you on a subject highly interesting to us both.—This is the expediency of publishing the Lectures of the late Prof. Peck. We have examined them carefully & corrected some slight errors in the expression, and innumerable ones in the chirography . . . This has enabled my Father & myself to form a judgement of their merits, which I wanted by leave to state with deference as well as candor.
. . .

These lectures . . . , as an illustration of the Linnaean System, are useful & pretty correct; . . . & [we?] conceive them to have been well adapted to the state of Natural Science in this section of our country, at the time when they were written. But while we admit all this, we must acknowledge with one of the committee, "that they contain nothing new," if we except the result of the Prof. own, individual observations on some very few species.[73]

About this time, Harris also was preparing for a break with Peck's instructions regarding descriptive conventions in entomology, as well as questioning his classificatory orientation. Writing to Thomas Say in November 1824, he observed that Peck had instructed him always to define species in Latin, a procedure that he had generally followed "though it savours somewhat of pedantry."[74]

It was during the 1820s, when he was living in Milton, Massachusetts, and engaged in the practice of medicine, that Harris did most of his entomological collecting.[75] But he was alone and, though he came to see Peck's limitations, he also lamented that his friend and mentor's death had taken away his one dependable source of information on entomological matters.[76] From that loss he reached out to the larger entomological community for aid, a move that set him on a course that transformed his personal interests in natural history, with their complex roots in his family and relations with Peck and others, to a career as contributor to entomological knowledge in all its facets.

Harris's first publication, on the saltmarsh caterpillar, has been referred to above. He began there with a statement on the importance of hay, and went on to explain that "The object of this paper is to attempt to elucidate the natural history of the . . . [insects], with the hope that it may lead to some sure method of exterminating them, or of limiting their ravages to a shorter period." He followed Peck in naming the insect and gave the description in Latin,[77] again according to Peck's preference. L. O. Howard, in his history of economic entomology in the United States, called it "a good, plain, practical paper."[78] Harris's second publication was "Description of Four Native Species of the Genus Cantharis," *Boston Journal of Philos-*

ophy and the Arts 1, no. 5 (January 1824): 494–502; all of the insects had been described previously,[79] but the paper was of sufficient interest that it was also printed in the *New England Journal of Medicine and Surgery,* to which it was initially submitted, in the same year. In 1826, Harris published the first of a number of articles in the *New England Farmer.* Between 1823 and 1830, he published twenty-seven papers[80] of which twenty-one first appeared in the *New England Farmer* (another three were reprinted there from elsewhere). The *Farmer,* therefore, was crucial to him in getting his research before the public.

The *New England Farmer* is not the type of publishing outlet one expects for scientific work in natural history even in Harris's period, and his early dependence on the agricultural press says much about the origins and initial orientation of his entomological work. It also reflects the state of science studies in New England in the early years of the nineteenth century. There was an apparent lack of interest in natural history there during the period, one reason being the perception that such pursuits had little practical value. Historian John C. Greene sees the establishment of the professorship and botanical garden at Harvard, to serve the needs of scientific agriculture, as one contemporary response to this objection.[81]

The formative years of Harris's scientific interest and career are bracketed by the rise and demise of the Linnaean Society of New England, a Boston-based organization originating in 1814. The society accumulated a notable collection of specimens (including insects), which for a time was well cared for but subsequently became a burden to the relatively few active participants. The Society essentially disbanded in 1823 when the initial energies and enthusiasm of the members dissipated, and regrettably an agreement by Harvard University to maintain and house the museum was not observed. The end of the Linnaean Society was a relatively brief hiatus, however, and was due perhaps more to organizational deficiencies than to any lack of inherent interest in the community. When the Boston Society of Natural History came into being at the end of the decade (1830), it more effectively tapped the various strands of interest and commitment developing in eastern Massachusetts.[82]

2

Career and Family

Harris never wandered very far from his place of birth. Even his travels seem to have been quite limited, the most distant being a trip to Philadelphia in 1822 to accompany an invalid patient. On that occasion he missed a chance to introduce himself to Thomas Say on account of the patient's condition, but he did have the opportunity to visit Charles Willson Peale's museum.[1] On the way home, he also visited the American museum in New York City,[2] apparently meaning the natural history museum established by John Scudder in 1811.[3] Though the excursion was for professional medical reasons, he did manage to do some entomological observing along the way, referring several years later to an insect that he had seen in abundance at Hoboken, New Jersey, on May 21, 1822.[4] Harris's other known trips from home were infrequent and confined to points in New England, and during his later years trips to the White Mountains of New Hampshire played a particularly important role in his life.

Harris initially practiced medicine with Dr. Holbrook in Milton after completing his studies, but in fall of 1823 he established an office in Dorchester Village (near Milton Lower Falls).[5] Harris was married in the home of his father-in-law, Dr. Holbrook, in Milton, where he is said to have lived while studying with the doctor and for several years thereafter. After his marriage to Catherine Holbrook in 1824, the couple resided in the House of the Suffolk Resolves in Milton, to which Harris had moved several months earlier.[6] Belonging to his wife's family, the house was the place of a famous meeting in September 1774, presided over by General Joseph Warren, which issued an assertive resolution against British oppression that subsequently was adopted by the First Continental Congress.[7]

In December 1909, entomologists meeting in Boston, along with representatives of the Milton Historical Society and the Science

Club of Milton Academy, dedicated a marble tablet at the Suffolk Resolves house, written by Harris's biographer Thomas Wentworth Higginson. It read,

IN THIS HOUSE FROM 1824 TO 1831 DWELT
THADDEUS WILLIAM HARRIS M.D.
BOTANIST, ENTOMOLOGIST; AND FINALLY
LIBRARIAN OF HARVARD COLLEGE

IN EACH CAPACITY HE WON
FOR HIMSELF FAME AND GRATITUDE

HE HAD THE MODESTY AND UNSELFISHESS
OF TRUE SCIENCE
WITH WHAT MAY RIGHTLY BE CALLED
ITS CHIVALRY OF SPIRIT[8]

WORK, STUDY, AND DISCONTENT

The sense of place that Harris manifested from his early life does not have a parallel in his occupational loyalties. They were, it seems, more a means to an end. Though he practiced medicine for more than ten years, relatively little is known of the details of that part of his life. His practice seems to have extended over a wide area attended to on horseback.[9] Harris made relatively few references to his medical practice in his correspondence and those were generally negative or lacking in enthusiasm. In an obituary notice of Harris, probably written by his Harvard classmate John G. Palfrey, he is described as an informed and skillful physician but lacking in self-confidence.[10] By 1828, Harris had come to confess that he was "thoroughly tired" of medical practice and "long[ed] to resign it for pursuits more congenial to my tastes" and planned to leave what he called "this fatiguing & disgusting profession" as soon as he reached an annual income of six or seven hundred dollars.[11] Two years later, he was further from that goal than before because of personal illness and the consequent neglect of his practice, which brought increased competition from other physicians. It was a situation from which he was not sure he could rebound.[12]

While Harris's years as a physician in Milton were not to his liking, it was the period of his most active entomological collecting and field study.[13] While in Milton, he was able, to some degree, to combine his medical rounds as a country physician, which he later described for a young colleague: "I was then . . . a country doctor, riding my daily rounds in Milton & the adjoining towns, among farmers & mechanics,—though a very rural, &, in parts, rough & sequestered district. My business often carried me on horseback through mere paths in the woods—where, when not hurried, I sometimes stopped to catch insects. In this way, carrying my specimens in pill-boxes &c, I made some of my most interesting catches."[14] His time available for entomological work fluctuated, so that in the winter of 1825–26, he was able to devote a good deal of effort to his studies.[15] Overall, however, scientific work was done only as time allowed, his entomological correspondence, for example, taking place over several days during "any moment of leisure."[16] The lack of time affected the character of his research work as well. In 1828, he explained to his friend Levi Leonard, "Often have I been called from witnessing some important change in my insects, or from extricating the description of some specimen, & been obliged to leave the pursuit for hours & even days. Often have my larvae been lost by starvation, & the pupae desiccated for want of moisture, because I was unavoidably detained from home."[17] Although the years in Milton turned out to be among his most entomologically productive, ironically, the late 1820s were described, at the time, as "the winter of my discontent."[18]

As early as 1826, Harris was casting his eyes in other directions, including the librarianship of Harvard University. His father attempted to get the appointment for him, but his efforts were too late and the position went to Benjamin Peirce (father of the mathematician and astronomer).[19] In his search for a new occupation, Harris also looked to the professorship and botanical curatorship once held by his mentor William Dandridge Peck (to be discussed below). He must have shared this sentiment with his friend Nicholas M. Hentz some time before 1827, because that year Hentz wrote, "I long to see you occupy that beautiful white cottage in the Botanic Garden in Cambridge, and to direct my letters to you with the title which you put on mine. I am pretty sure it would suit you better to sit in Prof. Peck's chair than to mend broken legs and make people swallow mercury and emetics and opium."[20] The next year the two friends considered the possibility of Harris leaving New England and going south (Hentz was then professor of modern languages and belles lettres at the University of North Carolina). The contemplated move was motivated by Harris's dissatisfaction not only with medicine but also with "this pernicious climate." He explained, however, that so long as he was in medical practice, he was compelled to stay where he was. The question of his teaching in a medical school also was brought up but he felt he was unqualified for such a position.[21]

Hentz continued to explore ways in which he could help his friend to gain a satisfactory position. In early 1829, he suggested that Harris come to North Carolina and join him in planting a vineyard with the expectation that the wine business there was about to develop substantially. Harris replied that he would consider the possibility and would make some inquiries among friends, though he was skeptical about what they would do to sustain themselves while the vineyard developed.[22] Hentz's own plans were in a state of uncertainty during this period. Shortly after the above exchange, he informed Harris that his professorship at North Carolina was in jeopardy and that there was a possibility of a move to the University of Alabama. Thinking again of his friend, he asked whether Harris would have any interest in a medical professorship if one developed, noting that "It is true the climate is hot, but what fine bugs could be caught there."[23] Harris's reply was negative, both for himself and for Hentz, observing that the Alabama climate was such that "miasma . . . poisons the atmosphere." As for the idea of a medical chair, Harris again said he was unqualified for such an appointment. But if the Alabama climate were surmountable he felt that, with preparation, he could teach obstetrics or materia medica. For the former, he was able to draw on ten years' experience, and for the latter his "knowledge of Botany and necessary acquaintance with the manipulation of drugs" would make it possible to qualify in a relatively short time.[24] Other naturalist friends also offered him encouragement to leave his situation and relocate—his friend Charles Pickering hoped he would come to Philadelphia where, "with mutual cooperation, we might contrive to make entomology flourish in this country."[25]

Though other professional appointments were considered, the great lure of his adult life was a professorship in natural history at Harvard. It was not an empty aspiration. Professor Peck had favored him by entrusting the younger man with posthumous publication of his manuscripts. Furthermore, President Kirkland was a friend of his father and appeared to hold Harris in high regard as did John Lowell, as discussed above. Therefore, when Peck died, Harris undertook to prepare a course of lectures in zoology with the expectation that he would be named as Peck's successor.[26] This was not to happen. Peck died on October 3, 1822, and by early November the Board of Visitors of the Massachusetts Professorship and the Harvard Corporation agreed to the appointment of Thomas Nuttall as curator of the Botanical Garden. There were insufficient funds to confer the Massachusetts Professorship on him, but he was appointed as lecturer in natural history.[27] During the summer and fall, Nuttall had been lecturing in botany at Yale and elsewhere and collecting minerals in western New England. His modern biographer, however, was unable to reconstruct the exact circumstances of his Harvard appointment, although various candidates were discussed in the press, and the most likely successor to Peck was considered to be Jacob Bigelow, then Rumford Professor and professor of materia medica.[28]

Harris's expectations of the curatorship and chair were later rekindled. As he recalled the events, "While the late Dr. Kirkland was President of this University, during an unexpected & prolonged absence of the excentric [sic] Mr. Nuttall from his post here, I was invited by the President to an interview, in which he expressed, in strong terms, his wish that I would accept the office of Curator of the Botanic Garden & Instructor in Natural History in case that Mr. Nuttall did not soon return. This, however, happened shortly afterwards, & Pres. Kirkland's invitation only remained as a testimony of his esteem & good wishes." This probably was on the occasion of Nuttall's return to England from late 1823 to June 1824,[29] for Harris goes on to say that, when Nuttall finally left Harvard "ten or more years later" (Nuttall left in 1834), Nathaniel Bowditch, as a member of the Harvard Corporation, promised him the new Fisher chair in natural history "so soon as the funds . . . would yield an adequate salary to a professor."[30] During the 1820s, Harris's connection with the university was in a rather more minor role than he had hoped for, as a member of the external committee to examine students in chemistry and natural history; his father also served on the same committee during this time period.[31]

LIBRARIANSHIP AND SCIENTIFIC ASPIRATIONS

Though Harris expected and desired the professorial and curatorial appointment from an early date, he was, as indicated above, also on the lookout for other opportunities. The librarianship at Harvard University was one such appointment, a post his father once held. Higginson said of him that "he seemed born with the librarian's instinct for alcoves and pamphlets and endless genealogies."[32] In reference to attempts by his friends to procure the position for him in 1826, Harris admitted that it was "a situation which would have suited me exactly."[33] The appointment finally came in 1831, following the death of Benjamin Peirce.

Nothing has been found in his own correspondence referring to the circumstances, but the disruption the appointment created in his life is attested by the fact that few letters survived from that year. When Harris gave up medicine for the librarianship of Harvard College in 1831, it required a relocation of his family as well as a significant career change for himself. Harris, of course, had spent four years in Cambridge as a student and therefore it was not a new environment. As had been the case in Milton, Harris's first home in Cambridge was family-owned, in this case by the Harris family rather than by that of his wife. Known as the Mason House, it was built before 1700 and was located on Dunster Street, just off Harvard Square. The house was destroyed by fire shortly after Harris and his family moved to a nearby location on Linden Street in 1839, but a reconstructed layout of the house and property has been published, based on the childhood memories of Harris's son Charles who was born there in 1832.[34] In family remembrance, "the garden became under our father's care a perfect bower of flowers. The only fruit tree was a pear-tree on the north side, bearing indifferent fruit."[35] Harris went to Cambridge in 1831 in advance of his family, beginning his new position in early September,[36] and his wife and children joined him the following spring. A note by his young son Thaddeus summarized the occasion: "Mary Allen and I came about May 1 / Emma, Cat, Hens &c / Came a few days after / Mother & Harriet came / Sunday, May 6." Mary Allen was a Protestant Irish servant, Emma and Harriet were Harris's young daughters.[37]

Harris's accomplishments as librarian are related in some detail in a later section, but an assessment of how the position fit into the structure of his life is relevant here. Higginson found that at the outset of his career as librarian, "the methodical and accurate habits of Dr. Harris promised to make the daily routine of duty agreeable."[38] The writer of an obituary, most likely his classmate John Gorham Palfrey, Unitarian minister, historian, and professor of sacred literature at Harvard from 1831–39,[39] evaluated Harris's career both in terms of what the university provided and what Harris both offered and sought: The librarian's

> office he administered twenty-five years, a much longer time than any of his predecessors. . . . By reason of the unfortunate poverty of that department of the institution in respect to pecuniary revenue, the scholarly function of presiding over its enlargement, and causing it to keep up with the improvements and demands of the time, scarcely belongs to the officer in charge of it. . . . Not an eminent bibliographer, with no controlling taste for books, he always found something in the library to be done, and in the library most of the day-light hours of all the week-days of the year found him employed.

Nature was the book which he loved much more to read than those upon his shelves; and that he could not devote himself to that study was the great cross of his life.[40]

Palfrey raises a crucial point for understanding Harris's life and career, namely, how he conceived and handled the necessity of work for financial support, and the relationship of that endeavor to the pursuit of his entomological interests. The idea of leisure is central to Harris's life concept, meaning available time to carry on his studies in natural history. As early as 1827, in a statement regarding natural history studies in the United States, he placed "leisure and inclination" as most important for success.[41] The significance of leisure was not particular to Harris; when his naturalist friend Charles Pickering heard of his appointment as librarian, he not only thought it would be "more congenial to your pursuits than the practice of medicine" but hoped it also would free Harris to come to Philadelphia during vacations.[42] For Harris, it was the promise of free time to pursue personal research interests that made both the librarianship and the potential of a professorship particularly attractive.[43] In an address in 1840, Harris generalized about the allures of natural history as the framework for a life and revealed his own fondest wish: "If the engrossing occupations & cares of life still leave us some moments for relaxation—happy are we if taste & inclination lead us to devote them to the study of nature; and happier still are they who can give their undividivided [sic] attention, & the whole of their energies to this pleasing task."[44] Time freed for study was the common element in Harris's search for a compatible career niche, not, it seems, the idea of an inherent relation between work for financial gain and contribution to knowledge. In this sense, his professional vision was conceptually foreshortened—that is, he put the individual at the center of the effort rather than the attainment of a socially sanctioned and supported position that would combine financial income and intellectual product.

Harris's son observed that to the library "were given conscientiously and honestly the hours of daylight necessary to the performance of his official work; to the others [science and antiquarian studies] he brought all the time that could be wrested from sleep and recreation."[45] By the mid-1830s, Harris realized there was less time available for entomology as librarian than there had been when he was practicing medicine.[46] In addition to his prescribed duties, he also had to work extra to supplement his librarian's salary, commenting in 1836, "the ordinary duties of my office command the greater part of my time, leaving but little more than is absolutely

necessary for relaxation, exercise, and attention to my family concerns." Additional work could only be performed in the early morning in spring and fall, "or during the long evenings of winter when
such extra service, if performed at all, must at much inconvenience
be done at my own house."[47] This explanation, in the official report
of the library, makes no reference to his work in entomology, unless
it is hidden under the category of "relaxation."

He wrote in 1842 to American ex-patriot physician and botanist
Francis Boott, then living in England, about the long delay in his
attempt to assemble an insect collection for exchange with British
entomologist John Curtis. Harris's account reveals something of the
way in which his day and life were constructed and the increasing
encroachment of his library and other duties on his entomological
work. The effort began in 1839, and "During the following summer
vacation (beginning on the 20th July 1839), I got together the insects which I now propose to send to you [Boott acting as intermedi-
ary], and wrote the first page of the catalogue. Before I could finish
the comparison of these insects & the determination of their names,
the vacation ended, & my regular duties in the University began.
From that time till the present winter, I have not been able to finish
the examination of the insects & the writing of the catalogue; having
been as fully occupied with other pressing duties during the vacations as in term-time. You can readily imagine that my office as librarian is no sinecure," requiring his services six hours a day
Monday through Friday. In addition to library duties he also was lecturing on natural history in the college.[48]

The lack of free time not only limited the effort he could devote
to entomology, it also tended to determine the character of that
work and his relations with other naturalists. In 1837 he wrote to his
friend Charles Zimmerman that although he was not at the time
able to make a full exchange for insects sent, "the time may come,
when I shall no longer be so closely confined to one spot, & so much
absorbed with other duties; & when that time does come, I mean to
go forth into the neglected parts of this & neighboring states, & collect largely of the insect treasures contained in them."[49]

Harris's friend Hentz reported a similar situation; his lack of leisure (by the late 1830s he was teaching school, with his wife, in Alabama) meant he no longer was able "to study any branch of
Entomology save the Araneides [a group of spiders]."[50] Harris, in
fact, recognized the general problem among American naturalists.
Writing to a European correspondent in 1836, he argued, "Other
pursuits necessarily form the business of our lives, and our duties to
our families & society leave us little time to devote to any branch of

natural history. The paucity of laborers in Entomology, & the engrossing nature of our other occupations oblige us to pay dearly for the time which we may devote to collecting & studying insects." He further explained that because of this difficulty, Americans place a "high value" on their collections and consequently may not be able to exchange one-for-one.[51] Harris conceptualized the situation in Europe as different from that faced by American naturalists. "In Germany & France the study of insects is provided for at the public expense, professors are maintained, & schools are established by means of which this may be pursued without difficulty. But in this country no establishment of the kind now exists." He did mention, however, that the foundational statutes for the Massachusetts Professorship of Natural History had required instruction on insects because of their relationship to plants.[52]

T. W. Higginson's memoir indicates that some of Harris's friends considered the position of librarian not as an end in itself, but a step toward assumption of a professorship in natural history, although Higginson thought otherwise—that is, that the library post had an actual and not simply an ulterior appeal to him.[53] Supportive of Higginson's own view is Harris's apparent reaction to the immediate search for a successor to Nuttall in 1834. He reported to Augustus Addison Gould, one of his close natural history colleagues in Boston, that friends of Francis Boott were attempting to procure an appointment for him as professor of natural history, "and there is a strong probability of their success." Commenting on the development, he observed that "although I disapprove of the principle for the monied men of the city to have the filling of our professorships at their pleasure, yet I believe that the choice in Boott's case will fall upon a man of merit & talent, who will prove a valuable accession to the [Boston] Natural History Society."[54] In fact Boott was elected to the Fisher chair in 1834, but he declined[55] and funds subscribed for support of the appointment were returned to the donors.[56]

The question of whether Harris thought from the outset that the librarianship was only a means to a professorship cannot be entirely settled by his own words. More than likely, in the early years of his library career he did expect to have enough leisure time for his insect studies and therefore the idea of the professorship was set aside as not necessary to his program for the study and promotion of entomology. By the mid-1830s, this assumption began to pass. Part of the reason has to be the development of his public career in science, especially the two editions of a catalogue of insects that he prepared for Edward Hitchcock's reports on Massachusetts geology and natural history (in 1833 and 1835).

By 1838, Harris had been appointed to prepare a report on insects for the Massachusetts Commission on the Zoological and Botanical Survey and was negotiating the preparation of an entomological report for the New York State Survey.[57] When the question of a professorial appointment in natural history at Harvard reemerged in 1838 (a motion on the subject apparently was introduced at the Boston Society of Natural History),[58] Harris was interested. In aspiring to the professorship, however, he realized his limitations and did not want to misrepresent himself as a possible candidate. In correspondence with his friend David H. Storer, Harris explored his prospects, especially the question of how he was perceived as a well-rounded naturalist, and in particular as to his botanical knowledge.[59] Admitting his relative weaknesses as well as his strengths, his interest was an enthusiastic one: "If a first-rate botanist be wanted in that office, the corporation must look further for a suitable candidate. But if a teacher of the elements of Natural History is required, who, by zeal & ardor in the department assigned to him, might eventually make up any deficiency in his previous qualifications—then should I, without hesitation, be willing to be brought forward as a candidate."[60] No appointment was made at this time, and Harris's situation continued to decline as official responsibilities in the library and other commitments accumulated, including his non-professorial employment as lecturer on natural history in the college.

Teaching Natural History in Harvard University

Harris's teaching in Harvard College began in the spring of 1834. For the next two years (1835 and 1836), Augustus Addison Gould filled the position of lecturer and Harris took over again in 1837 and continued until 1842.[61] The extra income he got for teaching was one of his chief motives for undertaking the assignment. In 1836, for example, he wrote to Gould to inquire whether he intended to teach that year, and if not, "the compensation might be an inducement" to him to accept if offered.[62] In 1838, after he began his teaching duties, he explained to a correspondent that circumstances required that he teach.[63] When his teaching career ended in 1842 with the appointment of Asa Gray to the Fisher Professorship, one of the consequences was the termination of "a small sum, in addition to my salary as librarian, [which] has been paid to me for the service. As both were not more than enough to maintain my family, the loss of the additional compensation is a source of considerable uneasiness to us, and will subject us to some privations."[64]

Although teaching brought some additional income, it also drained away precious time. In 1840, for example, he was teaching two courses, to freshmen and seniors, with six meetings each week, including recitations and "explanatory lectures." Although botanical lectures were voluntary, he had thirty students in the class. Each lecture took about two hours to put the sketches on the blackboard, in addition to other time required for preparation to teach,[65] and he had to gather botanical specimens on his own for use in the classroom.[66] For a textbook, he was "obliged to use Smellie's Philosophy of Natural History for want of a better—but the American edition has a pretty good summary of Zoology prefixed to it by Dr. John Ware." This is the work that Thomas Nuttall had used before him[67] and was, in fact, the standard text in the colleges at the time.[68] He looked about for a possible substitute, however, asking his friend Doubleday about natural history teaching in England and about the textbooks used.[69] It is apparent, therefore, that even if his motive for teaching was more practical than pedagogical, he did take the task seriously.[70]

Dupree, in his biography of Asa Gray, refers to Harris's recitations as "extreme examples of a drillmaster technique,"[71] as the foregoing might suggest. But Dupree also noted, as have others writing on Harris, that his enthusiasm and interest in his students helped to overcome this regimentation that, at any rate, was built into the teaching methodology of the time. Harris himself recognized the constraints of the system and sympathized with his students. In his first lecture in 1837, he pointed out that, with the short time available, the lessons from Smellie would, of necessity, be rather long. He was aware that the students would prefer lectures, but the College lacked illustrations or museum objects to support such an approach. He regretted that the students' course of study was so filled with modern and ancient languages, mathematics, and natural and intellectual philosophy that there was insufficient time for natural history. "The study of words forms a very large proportion of your academical pursuits—But 'words are the daughters of earth, things the sons of heaven.'"[72]

His supplementary lectures to seniors in 1838–39 included sessions on distinctions between animals and plants; anatomy of the human frame; comparative size of parts, and on the limbs and teeth of animals; the anatomy and classification of birds; fish; invertebrate forms, and on insect instinct; the senses of smell, taste, hearing (four lectures on the latter), and sight ("with diagram"); OuranOutang and chimpanzee; albinos; three lectures on migration (including migration of insects); and torpidity of animals.[73] His botanical lectures, as an elective, were popular, attracting from one-third to one-half of the class.[74] He also formed a voluntary evening class.[75]

T. W. Higginson, his nineteenth-century biographer, contributed his own remembrances of Harris as a teacher:

> I was fortunate enough to be among his pupils. There were exercises twice a week which included recitations in "Smellie's Philosophy of Natural History," with occasional elucidations and familiar lectures by Dr. Harris. There were also special lectures on Botany. . . . Even these scanty lessons were, if I rightly remember, a voluntary affair. . . . Still they proved so interesting that Dr. Harris formed, in addition, a private class in entomology, to which I also belonged. It included about a dozen young men from different college classes, who met on one evening of every week at the room where our teacher kept his cabinet, in Massachusetts Hall. These were delightful exercises, according to my recollection, though we never got beyond the Coleoptera [beetles].[76]

Furthermore, a Harvard Natural History Society was formed among the students, with faculty approval on May 7, 1837,[77] and Harris was one of its supporters.[78]

Teaching forced Harris to consider natural history overall and how his own interests fitted into the various aspects of the subject. At the outset, he was quite clear about the direction of his own work and inclination, explaining to Gould in 1836, when they were discussing which one would teach the course the next year, that "I have no wish to engage in any branch of Nat. Hist. except Entomology, meaning, if possible, to devote my future leisure exclusively to it."[79] During the course of his career, however, Harris used disciplinary affiliation and specialization—or lack of it—as a strategic factor so that his professions in regard to subject matter were not always consistent, even when his practice was.

His initial interests were botanical and even in 1823, at which time his interests in entomology were well formulated and his first paper on the subject appeared, he was not well acquainted with other zoological groups such as fishes and reptiles.[80] To his correspondent Thomas Say, equally noted as a conchologist and entomologist, Harris confessed that he had "no decided taste for shells."[81] When he was passed over for appointment at Harvard in 1822—when the position of curator of the Botanical Garden and lectureship in natural history went to Nuttall—he vowed to give up all botanical and zoological studies except for insects, "which seemed to offer me many attractions, & which I thought was then the only branch of natural history in which I could expect to become distinguished." Harris seems to have remained committed to his resolution on specialization, though at some point before 1838, he "returned for a short time to botany" when he delivered two lectures on the subject for the newly established Dorchester Lyceum.[82] Having made his decision to confine his work to entomology, on subsequent occasions, when the possibility of a professorship reemerged—especially in 1838 and in 1842—he also tried hard to show that he was not unacquainted with, and certainly interested in, botany. That was his point in referring to his lyceum lectures mentioned above, to show that he had not entirely abandoned botany. Against Gray in 1842 he both noted his own experience in teaching botany and zoology and Gray's ignorance of zoology, the branch that was the favorite of the students in the college.[83]

RESOURCES FOR SCIENCE

In order to prosecute his specialized studies, both specimens and books were required, in addition to apparatus or equipment peculiar to entomology. The search for the wherewithal for research involved both his personal resources and those that were provided by the community at large. Harris realized at the outset of his entomological career that access to resources was going to be a central concern. In 1827, he wrote one of his most remarkable letters, addressed to one of Boston's important patrons of science, John Lowell (1769–1840), referring to Lowell's interest "in the promotion of the Natural Sciences" and "giving a brief view of their present state in this country, & . . . pointing out what appear to be the objects most deserving of attention." Harris noted that, next to an interest in the subject itself and the time to pursue it, "nothing is more essential than collections of objects, & books for consultation."[84]

In 1823, when he took an initial step into the larger natural history community by writing to Thomas Say, Harris's personal collection of insects included a little more than 500 species[85] and in the next four years he added some 700 more.[86] By 1836, he was able to report that his personal collection was likely one of the best in the United States,[87] while by 1842 he did not equivocate in calling his collection "not only the best, but the only general one of North American insects in this country."[88] Through a persistent effort, Harris's collection had developed through a variety of means. As indicated above, most of his active insect collecting took place before he began working at Harvard in 1831. Thereafter, he had to depend chiefly on exchanges or on the collecting efforts of others. By various means, Harris developed a significant collection, which was bought by the Boston Society of Natural History after his death. In the society's account of the collection, it was noted that the accumulation

> of native insects in Dr. Harris's cabinet has a peculiar value as containing many typical specimens of species described by himself, Say, and other naturalists, and also from its completeness in all its parts. It contains –

4838	specimens	of	2241	species	of	Coleoptera.
181	"	"	76	"	"	Orthoptera.
620	"	of about	300	"	"	Hemiptera.
267	"	of	146	"	"	Neuroptera.
1125	"	"	602	"	"	Hymenoptera.
1931	"	"	900	"	"	Lepidoptera.
796	"	"	395	"	"	Diptera.

In all, the collection had 9,758 specimens of 4,660 species, not including a large number of specimens not classified and was described as "in good condition."[89]

A noted entomologist of the next generation, Samuel H. Scudder, who had edited Harris's correspondence and other papers,[90] wrote that "The very existence of his cabinet, with the vigilant care which must be taken for its protection from insect pests, in the open drawers which alone his straightened circumstance allowed him, must have consumed every moment that he could spare from his official duties."[91] The maintenance of the collection as well as the availability of the tools for its study was indeed a concern. After he received Say's collection in 1836, he wrote to Mrs. Say about its condition and noted that his insects never had suffered the infestations that characterized her husband's, which he credited "chiefly to my having kept them in close [sic] drawers, lined with cork, & impregnated with camphor; and to my having given them a baking every spring."[92]

Over the years, Harris made various references to magnifying devices that he had available or that he found desirable. He was imaginative in his approach to the apparatus to be used in his studies, in addition to accepting the use of simple devices with which he could become adept. In an undated letter to Rev. John Prince, a noted instrument maker who produced apparatus for Harvard as well as for other colleges, Harris described his desire for an apparatus to aid in his drawings of insects which also reveals something of his working habits: "You may recollect that I expressed a desire to have some instrument by which I could obtain correct outlines of insects of the natural size. The following contrivance has occurred to me, & I think would answer the purpose if I had your aid in perfecting it," adding, "It is essential that my drawings shd be made by candle-light, & it is also essential that the paper on which the drawings are made should be on a horizontal surface, because I use a crow quill, & the ink will not flow readily in a pen if the surface to be drawn upon be not horizontal." Harris continues,

> If you can provide a lucernal microscope that will answer the purposes of making drawings of insects of the natural size, & enlarged from 1/4 to 5 times that size, will you please inform me what would be the lowest price at which it could be afforded.
>
> Can large insects be exhibited in these microscopes as well as small ones? Sometimes I should wish to draw insects whose actual size would be from one inch to 3 or 4 inches, & which therefore it would be unnecessary to magnify.[93]

It is not known whether Harris ever procured the described device.

The drain on his resources is more obvious in his frequent and ongoing efforts to procure the books he needed for his research, most of them by European authors. He did not equivocate in placing a higher value on European books than on European insects, and in the political economy of resource development he sometimes offered American insects for European books.[94] Harris's library was purchased after his death by John P. Cushing of Watertown, Massachusetts, who also was the largest donor for the purchase of his entomological collection and manuscripts, and the library also went to the Boston Society of Natural History. The book collection included some three hundred volumes and pamphlets,[95] but in the static form in which it was enumerated in the *Proceedings* of the society,[96] nothing is told of the effort and at times despair and sacrifice by which the collection was assembled. The problem related not only to his interests in description and taxonomy but also to the agricultural aspects of his entomological studies as well. For example, in the late 1840s, he received volumes of the *Prairie Farmer,* published at Chicago, in exchange for his own entomological publications, but in order to receive it in the future he proposed writing occasionally for the journal, explaining that "several other Agricultural papers are sent to me on these terms."[97]

Harris commented to his friend David H. Storer in 1836 that the lack of funds to buy books was always a problem that "rather increased, or appeared of more importance, as my knowledge of species was enlarged, and I soon found myself in possession of a very large number of insects which could not, with any propriety, be arranged in any of the genera described in my books."[98] In the late 1830s, as his interests turned increasingly to the Lepidoptera, he found the need for more specialized works and in 1838 reported to a correspondent that "I have recently imported from Europe some very valuable works on this order of insects, at an expense of near 500 dollars."[99] A year earlier, he had told a correspondent that he was not able to afford the "large expensive" European works on Lepidoptera and Diptera.[100] It is not determined whether any of the books that Harris imported at this time fit the "large expensive" category, but it is clear that he had determined to accumulate the essential works for his entomological writing. In April 1838, he told John Eatton LeConte that, although he had not been able to afford the publication by LeConte and the French entomologist J. A. Boisduval on American Lepidoptera, he had hopes that labor he anticipated doing for the New York Natural History Survey would give him the necessary funds.[101]

Although he apparently had established good connections in Europe for the ordering of books (probably aided by his work in the library), and felt for a time that he could afford them, in late 1840 he had to write to the firm of Hector Bossange & Co. in Paris that "Circumstances render it expedient for me now to close my accounts, & to contract no further debts with your house till the balance due to you is settled. I am resolved no longer to have a running account (comte courant) with any foreign house . . . From & after the receipt of the present letter you will observe that my orders for books are suspended, and you are hereby requested not to send me any more continuations (suites) until they are ordered anew by me." Harris says his reason for this change in course was the significant losses, to the amount of more than sixteen dollars, that he suffered because of the rate of exchange, which "make the books come very dear, & I must try to avoid them in future."[102]

In spite of his efforts Harris always had to struggle to find the required literary resources as support for his entomological studies, and this was particularly evident as he became more interested in the study of the Lepidoptera. He published a statement of the need for more adequate support and its consequences for public science in the two editions of the catalogue of insects that he prepared for Edward Hitchcock's report on Massachusetts geology and natural history in the mid-1830s. There he wrote that, in preparing the inventory of Massachusetts insects,

I have had before me not one of the numerous catalogues of insects which have been published in Europe. Many of the most valuable, important, and standard works on Entomology are known to me only by name! they are beyond the means of an individual to procure, and are not to be found in our public libraries, which are lamentably deficient in the most approved works on this subject. In the present state of the science, entomology cannot advantageously be pursued without books, and collections from other countries for study and comparison: for the want of these the second edition of this catalogue, though much enlarged, is very far from being complete; a great number of our insects have necessarily been left without names, it being impossible, without consulting authorities inaccessible to me, to refer them even to their proper genera. This is more particularly the case with the Lepidoptera, the list of which is now reprinted without alteration from the first copy.[103]

In 1838, he confessed to British entomologist John Curtis that "a small income & a large family" prevented him from buying many required natural history books, a situation that led him to seek a named collection of European nocturnal Lepidoptera. In this regard Harris was pleased that such a resource could be procured through an exchange of insects with Curtis.[104]

Harris did not entirely depend on sources he could afford to buy to support his research needs. He spent a good amount of time copying from and indexing various entomological works as an aid to study and classification. The result is effectively described by his former student and early biographer, Thomas W. Higginson, who refers to the papers of Harris as they were deposited after his death with the Boston Society of Natural History, where it became

apparent by what vast labor Dr. Harris had compiled for himself the literary apparatus of his scientific study. A mass of manuscript books, systematized with French method, but written in the clearest of English handwritings, show how he opened his way through the mighty maze of authorities. First comes, for instance, a complete systematic index to the butterflies described by Godart and Latreille, in the Encyclopedie Methodique. Every genus or species is noted, with authority, reference and synonymes [sic]; the notes being then rearranged alphabetically and pasted into a volume—perhaps three thousand titles in all. This was done in 1835.

Then comes a similar compilation of the Coleoptera from Olivier; twenty foolscap pages, giving genus, species, locality, and even measurements, to the fraction of an inch. Then there are three manuscript volumes containing an index to the four volumes of Cramer's "Papillons Exotiques"; one devoted to Stoll's "Supplement," and two to Hubner's

"Exotische Schmetterlinge." For Drury's "Illustrations of Natural History" there are two of these elaborate indexes, made at different periods; one based on the original edition in 1770–3, and the other on Westwood's reprint of 1837. So beautifully executed is all this laborious work, that it is still as easily accessible as print, though the earlier sheets are yellow and torn.[105]

None of the works mentioned in Higginson's summary is in Harris's library as deposited with the Boston society, and it appears that he must have borrowed the works from a public institution or from private collections in order to prepare his compiled indexes.

Harris's efforts to assemble the resources for his entomological work could not overcome the other pressures and deprivations of his life after the events of 1841–42 described below. In 1845, with remorse, he observed both the success and failure of his efforts to create a viable situation for himself, having to abandon the study of natural history "just at the period when, from experience & from the apparatus of collections & works brought together by my own efforts during some twenty years of my life, I am better fitted and prepared to prosecute the study of insects (a branch of natural history almost wholly neglected by other persons in this country), than at any former period."[106] Substantially more titles published in the 1830s were in Harris's library than for any other decade (nearly twice the number that were published in the 1820s). While that does not confirm when they were purchased, it does reflect the most active part of his entomological career, from the mid-1830s to the early 1840s. It is suggestive of the fact that he was able to acquire many publications in a timely way as an aid to his research. In contrast, the number of publications published in the 1840s and 1850s showed a noticeable reduction and follows other evidence of a decline in his entomological activity after the early 1840s.[107]

It is a given in the discussion above regarding Harris's personal resources for research that the larger community also offered material support. The two were complementary, and sometimes Harris was a consumer of that larger resource and at other times he was a contributor. In the context of his own career, part of his strategy for advance was to engage the public (or public institutions) in the assemblage of research sources that served not only his own but their purposes. In 1826, Harris's friend Nicholas M. Hentz wondered why John Lowell did not entrust Harris with the William D. Peck collection that he had purchased.[108] At this time, Harris was preparing a list of entomological books for Lowell, who had promised to procure certain works for the Boston Athenaeum library where Harris already had access to other books of interest to him.[109] The Athenaeum was an important resource for him but he recognized that other subject interests there competed heavily with natural history.[110] In time, the library of the Boston Society of Natural History (founded 1830) became part of the equation of research in the region, and it was recognized when the society was founded that "the great difficulty in the way of making advances in the study of natural history, was the want of books."[111] Harris certainly would have agreed with this statement, and one of his contributions to the society library was his efforts to bring the insects and books of N. M. Hentz to Boston (see account below).

Beginning in 1831, Harris was directly involved professionally in developing the resources for natural history study in Boston, as Harvard librarian and later as a curator at the Boston Society of Natural History. In his first report as Harvard librarian in 1832, Harris referred to deficiencies in the library, noting that while "some valuable and even splendid and costly works on Natural History" are there, "the most useful, celebrated & essential works of moderate price and recent date" are not.[112] Harris took more than an official approach to the problem. In a letter to President Quincy on June 24, 1834, in which he presented his bill for services as substitute lecturer in natural history (filling in for the departed Thomas Nuttall), Harris explained why he had "put these services at a very low rate." The reason was that he wanted the Harvard Corporation to use the money not needed for Nuttall's salary as curator of the Botanical Garden to buy much needed natural history books for the library. But this was not entirely an act of altruism, for Harris explained that, if ordered soon some of the books could be received by winter, "a circumstance which would be particularly gratifying to me, inasmuch as I have engaged to correct & enlarge some of the catalogues inserted in Prof. Hitchcock's 'Survey of Massachusetts'."[113] In spite of Harris's plea and the sacrifice he made by claiming a lesser compensation for his lectures in the college, he found himself "mortified & disappointed at the coldness with which all my attempts to move the corporation to this effect have been received." On the occasion, Harris pledged to make one more concerted effort to persuade the university or the Boston Society of Natural History to procure the entomological works he needed. Otherwise, "I will dispose of my collections & books, & give up forever the study of Entomology."[114] He, of course, did not give up his entomological work, but instead (as discussed above) undertook to purchase on his own, and at personal sacrifice, many of the works that he needed.

In 1838, he again assessed the natural history collections (along with other areas of deficiency) in the University Library with the hope that some special effort would be made toward their development, in conjunction with the construction and move to a new library building. The librarian-naturalist stated his case to the Harvard Corporation in personal, institutional, and scientific terms:

> The deep personal interest felt by me in the subject of <u>Natural History, in all its branches</u>, the constantly increasing attention which is paid to this science in this country, so rich in unknown & imperfectly described animals & plants, and the consequent demands made upon the library, for works treating on this subject, by persons who are desirous of engaging in <u>a science which is not to be mastered without the aid of books</u>, and, lastly, the recent changes in the course of studies in the university, which give to students more time for pursuing Natural History, will, I hope, be deemed good & sufficient reasons why I should urge you, Gentlemen, to take immediate measures to make large & liberal additions to the department of Natural History in the Library.

He observed that, with a few exceptions (including the gift of Audubon's *Birds of America*), no resources in this area had been added since the death of Peck in 1822. He addressed the question of individual and institutional responsibilities as he characterized the need not only for manuals and common works, but also "the larger, voluminous, & more expensive works of these subjects" that are beyond the ability of individuals to procure, and "which most properly should have a place in a <u>public</u> library." To make the point more obvious, he mentioned that, not only does the Boston Athenaeum have many more works on natural history than does the University Library, but, "of standard writers on Botany, Zoology, & Mineralogy, there are in my own <u>little</u> private library more than 100 volumes which are not to be found in the Library of the University."[115] Though works of interest to Harris and to other naturalists undoubtedly were added to the University Library thereafter, and by the mid-1840s he had available better resources than ever before to carry out his entomological research, it is one of the sad ironies of Harris's career that, only six months before his death, he announced that the Library had received a bequest of $5,000 "to be applied entirely and with all convenient despatch to the purchase of books on Natural History."[116]

While Harris was involved with the Library's efforts to develop book resources for entomological study, he also was engaged, both on his own and in the institutional setting of the Boston Society of Natural History, in the development of public insect collections that complemented his own accumulation of specimens. Sometimes

these activities came together. For example, in 1838, he entered negotiations with Nicholas M. Hentz, then resident in Alabama, to collect Lepidoptera (butterflies and moths) for the Boston Society for
pay. These were the insects that Harris was then studying, so the society's purchases would benefit him directly. He went on to explain,
however, that "Larvae & pupae are not objects to make a show &
attract attention in the cabinet of our Society—& nothing has been
said about them; but I shall be glad to become a purchaser of these
on my own account, & on such terms as you may determine
upon."[117]

His temporary possession of the collection of Thomas Say was one
of the most important events of his entomological career. In 1833,
Harris sent a large collection of his insects, 1,970 specimens in all
and including some 1,800 species, to Thomas Say, then at New Harmony, Indiana, asking his help in identifying them as an aid to Harris's revision of his catalogue of Massachusetts insects for the
Hitchcock report.[118] Regrettably, Say died on October 10, 1834. Harris made some early inquiries about Say's library and collections,[119]
and within a year he received his entomological manuscripts from
Mrs. Say and planned a trip to Philadelphia to examine his insect
collection.[120] By the middle of the next year (1836) he had Say's insect collection in his possession, through arrangements made by
Charles Pickering with the Academy of Natural Sciences of Philadelphia, and with the understanding that Harris "should put them in
good order & return them in a condition to be preserved."[121] It was
anticipated that he would retain it for several years.[122] Shortly after
the collection was received, he reported to Charles Pickering that it
was infested with anthreni and many specimens were loose and broken and separated from their labels.[123] This fact, plus the deficiencies in the collection itself as they related to Harris's own needs,
deflated his expectations of substantial aid from the collection,[124] although by referring to what he calls "the relics of Mr. Say's collection . . . many of my doubts have been removed & much further
knowledge has been attained."[125] While Say's early death and Harris's work with his manuscripts and collections seems to have complicated his professional life, eventually the circumstances broadened
his entomological outlook.

Harris's effort was more institutionally grounded although, in the
short term, equally frustrated, in his attempt to procure the Hentz
Collection for the Boston Society of Natural History. This activity
was carried out at about the same time that he was engaged in the
attempt to salvage and use Say's collection. Initially, the Hentz Collection was offered to Harvard University at a cost of $1,600, but it

was turned down for reasons not known to Harris.[126] When Hentz said he would sell the collection in Europe, however, Harris undertook on his own to raise concern and the necessary funds for its purchase by the Boston society.[127] His efforts at a subscription generated relatively little interest, and Harris's circular sent to individuals reportedly susceptible to such a solicitation received no response[128] so he had to rely on acquaintances of these prospects to make personal appeals.[129] The money involved was $400 for books and $600 for the insects, and after six months of effort $565 had been pledged. At one point, he asserted that Hentz's insects would be bought even if he had to pay the balance himself, and tried to interest the society in sponsoring lectures to meet the expenses, offering to donate at least four lectures himself.[130]

Though ultimately successful, Harris was less than forgiving of the apparent lack of interest in the collection's purchase. Harris remarked at the Boston society's annual meeting

> I congratulate the society in the acquisition it has received, and although the tediously protracted negotiation has caused me much anxiety and vexation, and the small and lingering success which has attended my efforts in your behalf has subjected me to severe mortification and disappointment, I cannot but feel happy at the result. It is my hope that we shall have here in entomology, as well as in other depts., a standard collection rich in genera and species, as complete as possible in the productions of our own country, arranged and with the names affixed to every described species.[131]

It was indeed a substantial addition to the entomological resources in the Boston community, containing 14,126 specimens, 12,811 of which were American.[132] Harris understood that, when combined with its own collection, the entomological resources of the Boston Society of Natural History would be the best available in the country.[133] His obligations went beyond the procurement of the Hentz Collection to encompass curatorial responsibility on behalf of the society—he brought the collection to Cambridge as early as December 1836 so that he could examine and organize it for the society.[134] By 1838, with the Say and Hentz collections accessible to him, Harris was able to overcome some of the difficulties posed by the ever-present constraints hampering his efforts to be naturalist as well as librarian, and he was poised to engage in a more ambitious entomological program.[135] Circumstances largely outside his control would determine otherwise.

Events of 1842 and the Consequences

In Harris's life, all events flowed to and from the period around 1842. Within a year or so his historical reputation was assured with the publication of the *Report on the Insects of Massachusetts, Injuriou to Vegetation* (1841), he failed to achieve the professorship in natural history which went instead to Gray, and his duties in the library took on a more demanding character. As discussed in the next chapter, in the mid-1830s to the early 1840s, his entomological research career was in full flower. After 1842 and until the end of his life, he was able to do less and less work in natural history,[136] as "he yielded to the inevitable, and became more and more absorbed in his official [library] duties."[137] As his memoirist Higginson stated, he was forced by circumstances to the "final abandonment of the hopeless attempt to be librarian and naturalist at the same time."[138]

The degree to which Harris had become overcommitted, and his emotional and physical reaction to the stress, is painfully revealed in an April 1841 letter to Edward Doubleday:

> I tried to begin a letter to you, but was as often interrupted, & prevented from going on, by some demand on my time & services. . . . I have been greatly distressed & dispirited, & worn down in mind & body, & am still far from being well. Close confinement, & want of exercise, & continued mental exertion, as my medical friends think, brought on a state of mind which reduced me almost to despair. The suffering which I passed through in the month of February & beginning of March I cannot describe . . . The thoughts of labors unfinished, engagements not fulfilled, & duties pressing on me for immediate performance with the apprehension, whether real or imaginary, of having my labor suspended by sickness or death, kept me in a continued nervous agitation excitement, [*sic*] while wandering pains . . . disturbed all the natural functions of the System. My eyes too have suffered so it is painful to them to read small writing for more than a few minutes at a time.

As Harris indicates, the pressures were multiple—these included the planned move of the University Library to Gore Hall during the forthcoming summer, the completion of his report on Massachusetts insects, the need to prepare the collection of Thomas Say for return to the Academy of Natural Sciences, at their request, and the deliverance of his natural history lectures to the undergraduates.[139]

The cornerstone of the new library building (Gore Hall) was laid in 1838,[140] and the character of Harris's work as librarian began to change in that year when plans were floated to increase the size of the collections as well. Harris took some initiative in this regard by suggesting the need to identify weaknesses among the holdings and to begin solicitation of funds in a timely way so the books could be ordered from Europe, fearing that "there is reason to apprehend the new building may present a beggardly array of empty shelves, notwithstanding the crowded state of the present library."[141] In addition to these library duties, Harris was working on his state entomological report, which, he reported in December 1840, had been "nearly done for some weeks."[142] In April 1841, however, he reported that he still had 150 pages to write, even though the majority already had been printed.[143]

The request from the Academy of Natural Sciences of Philadelphia for the return of Thomas Say's collection came as the result of a vote of the Academy on January 12. It was a crushing and threatening blow, as Harris interpreted it in the context of the many tasks he had to do at the time. There was some confusion as to when he had received the collection. The Academy said 1828, whereas Harris reported that he had had it only since 1838.[144] In fact his contemporary correspondence indicates that the Say collection had come in 1836.[145] Although it provided Harris with an important resource and advantage in his entomological work, it also was a significant responsibility and a burden from the beginning. In desperation, he wrote in February 1841 to his friend Charles Zimmerman in South Carolina for advice and help:

> I am in great trouble, & you alone can help me out of it. The amount of the trouble is this—The Academy of Natural Sciences of Philadelphia have desired that Say's Collections should be returned with as little delay as possible. It would give me the greatest pleasure to be able to send back the insects immediately. But you know the ruinous condition in which most of them came to me & can form some idea of the trouble which they have given me, on account of the vermin with which they swarmed. I am absolutely in despair about them. I am pressed with a load of official & other duties which have hitherto prevented me from putting this collection into decent order, & I know not how to get through with this labor now, with all the other engagements which must be fulfilled during the coming spring & summer. So much has this requisition of the Philadelphia Academy harassed me that my spirits are in an extreme state of agitation & depression; my sleep is disturbed by anxious visions, & is without refreshment; & I sometimes think I shall become insane. With the thought of having so much to do, I can do nothing well; & to prepare these insects for immediate return is absolutely out of my power.[146]

Zimmerman did not receive his letter until June, at which point he was unable to come to his aid. But he advised Harris to return the collection as is to the Academy, or suggested that for a small fee he perhaps could buy it, based on Zimmerman's assessment of the lack of real interest in entomology among the Philadelphia academy members.[147] In response to further requests for return of the collection, Harris complied in January 1842, shipping "three large boxes or cases . . . containing a considerable & (as I presume) the most valuable part of the collection of insects belonging to the Academy, which has long been in my hands."[148]

In the same month that he returned the Say collection and thus removed that burden from his attention and conscience, Harris reported that he was more free for entomological studies than in the previous winter. He apparently was poised for a return to work he had underway on the Lepidoptera,[149] but in fact he was about to enter one of the great dark times of his life. The story of Harris's failure to achieve his long-held hope for a professorship in natural history at Harvard is well told in Hunter Dupree's biography of the man who won the chair. Asa Gray (1810–88) arrived on a first visit to Boston in January 1842 in the wake of a December letter to Benjamin D. Greene (president of the Boston Society of Natural History from its founding in 1830 until 1837),[150] to whom Gray had written about prospects at Harvard. During his visit, he stayed with Greene and met and dined with Harvard president Josiah Quincy. He also met with George Barrell Emerson, a botanist and then president of the Boston society, whom Dupree described as "the man who had decisive influence in selecting the professor." Direct discussions with Gray in regard to the appointment were authorized by the Harvard Corporation in late March, which amounted to an offer of the position if the terms were acceptable. President Quincy asked for an answer by the end of April.[151]

It is not clear when Harris first heard of these developments. In a letter in mid-February to Francis Boott, who had once been offered the Fisher Professorship and turned it down, he reiterated the various burdens that he carried, and pointed out that "I have performed all the duties of a Professor of Natural History in the University for the past several years past without, however, the honor of a regular appointment."[152] However, this letter relates primarily to the delayed but finally realized project to gather a selection of insects for transmittal to British entomologist John Curtis and suggests rather that Harris was getting back to the business of his entomological research and obligations after completion of the *Repor* and other events outlined above. Harris's letter of April 25 to his friend David H. Storer, in which he inquired about rumors of an appointment of Gray,[153] therefore, is most likely genuine in its sur prise, though it is apparent that he was aware that plans were afoot to fill the professorship. The news came to him at a time of mourning. Harris's father, after an illness of a week, died on April 3, 1842, and was buried four days later.[154] Their closeness and the father's support for his ambitions is revealed in Harris's later remark that he was grateful that his father was "spared the pain" of knowledge that the son was passed over in favor of Gray.[155]

The rumors that Harris heard came from the Boston Society of Natural History, where it was said that a society committee was considering a recommendation for the appointment and that the leading candidate was Gray. Harris also referred to rumors from the previous winter that anatomist Jeffries Wyman had been promoted as a candidate and that some of his friends, not inclined to Gray, were leaning toward Wyman. The reason for this possibility, as Harris stated his knowledge of the rumors, was that these friends did not think the Corporation would appoint him, in spite of his many years of service as lecturer in natural history. Harris could not understand this possibility and felt, as well, that the college faculty would favor him. Harris opens himself to his friend Storer, searching for enlightenment on the situation no matter how personally painful, asking Storer how his former students, "the young men who annually go forth from the University to cultivate science in other places, & to recruit the ranks of the Boston Society of Natural History," assessed him in regard to the professorship. But he also urged support for his cause, if support is there, and gave Storer the names of some friends and others whom he thought could promote his claim.[156] Whatever the particular attitude of the Corporation toward Harris as a naturalist, his attempt to rouse support was clearly too late. Harris's prospects for a professorship ended when the formal appointment of Gray was made on April 30.[157] While still coping with this loss, his personal and family sorrows continued when his father-in-law, Dr. Amos Holbrook, with whom he had studied and practiced medicine, died in June just ten weeks after his own father's demise, and with these deaths, "the pleasant homes of childhood have passed into other hands, and the places that once knew us will know us no more."[158] As a consequence of the several events of early 1842, Harris's mature years were cut off from those of his childhood and from the professional dreams derived from early manhood. He was then forty-six years old.

After the Gray appointment and in light of the way it was achieved, Harris confessed to Edward Doubleday that "It is mortifying to me to be obliged to remain here, but my salary as librarian is necessary for the support of my family. Time may reconcile me to my lot, . . . it can never bring back the hopes & motives for exertion that have heretofore sustained me."[159] Among the painful aspects of the Gray appointment for him, carried out in secret according to his rendering, and kept even from the members of the Boston Society of Natural History,[160] was what he considered the betrayal of his interests by Benjamin D. Greene and George B. Emerson,[161] former and incumbent presidents of the Boston society. This must have hurt his sense of worth as a naturalist to have been so judged by leaders of the Boston scientific community and contributed to the sapping of his energy and enthusiasm. Dupree says that Harris's "view of his own qualifications [for the professorship] was still soundly based on the old naturalist's ideal of a well-rounded teacher," but this conclusion uses his own words against him, from an argument prepared for the occasion in an attempt to elevate his chances for the appointment. In all likelihood, Harris was both too much the specialist and in the wrong specialty to have succeeded in any contest with Gray, a conclusion underscored by Dupree's point that Gray's ultimate advantage was in being a botanist.[162]

In the wake of the Gray appointment in 1842, Harris engaged in discussion of the event as a loss for zoology against botany as well as a personal loss for himself. He was encouraged by Pennsylvania zoologist Samuel S. Haldeman, who was in Cambridge before the appointment and had spoken out in Harris's favor. After reflecting on Gray's appointment, Haldeman reached the conclusion "that the chairs of 'natural history' in our colleges, are no service to zoology, neither tending to its advancement, nor affording working means to zoologists to pursue their studies." In fact, Haldeman had recently been made professor of zoology in the Franklin Institute, Philadelphia, and in his "introductory remarks" he contended that "our botany is immeasurably in advance of our zoology, & that if one must be neglected, it should be the former; whilst the latter should have every support possible." Haldeman went on to argue that there should be separate chairs for zoology and botany and, quoting from his own introductory lecture, justified it in part from the point of view of the need for professorial specialization in order to promote research: "In grasping at too much, he [the professor] may not have the time to make a single discovery, or original observation—to add a single chapter to the gross amount of knowledge upon the subjects he is expected to understand; and he consequently degenerates into the mere compiler."[163]

In reply to Haldeman's letter, Harris retold his long-held expectations for a professorial appointment and the Massachusetts Professorship in particular, "a chair originally & expressly established by members of our Agricultural Society, to promote the study of Entomology & Botany, kindred sciences, closely connected with Agriculture." Although this reference pointed to the special relation of insect and plant that played such an important part in his scientific work and interests, his focus still was on his specialization. His loyalties are revealed with the hope "that American entomology will not suffer by the defeat of my long cherished hopes, and that new votaries may meet with better success."[164]

There are no preserved copies of Harris's letters in his own papers for the year 1843, although it had long been his practice to retain the drafts of his correspondence for reference.[165] The reasons for this hiatus are revealed when he again took up his epistolary connections. In February 1845, he wrote to minister and entomologist John G. Morris: "My correspondence with you, & with other entomological friends, has been interrupted not from choice, neglect, or forgetfulness, but from necessity. From month to month, I have been expecting & hoping for some change in my favor, that would give me a reprieve from engrossing occupations, & allow me time to think of, & make suitable returns to, friends for favors received. You will better understand my position by a few explanations. An inflammation of the eyes, supposed to be rheumatic, first put a stop to my entomological pursuits, & left a permanent weakness of sight, that renders it painful & difficult for me to use a microscope except during the <u>day</u>, & in a <u>good</u> <u>light</u>." Reference to his physical afflictions is followed by a detailed description of his pressing duties in the University library. Furthermore, "In the winter evenings of 1843–4 I had to draft plans & obtain estimates for a house that has since been built for me, and the superintendence of the work, the business arrangements, and the time required to remove & get settled in the new habitation, added not a little to my trouble & duties. The settlement of my father's estate and the arrangement & disposition of his numerous papers & manuscripts has partly fallen on me also,—the papers having come into my hands as heir thereto—and these still take up most of my evening hours."[166]

The Harrises began to plan for a home of their own in the early 1840s, and the death of Catherine's father, Amos Holbrook, in June 1842 provided inherited means to undertake construction, on Holyoke Place,[167] which, like their other Cambridge residences, was located near the library, between Harvard Yard and the Charles River. They moved into the new home in November 1843.[168] Writing to his friend Edward Doubleday in February of 1844, on that occasion Harris took "some pleasure in describing the house, because the plan was chiefly my own, though revised by an experienced architect." He referred to it as a

> cottage 2 stories high,[169] . . . . made comfortable by convenient arrange-
> ments, with accomodations [*sic*] for our friends. I wish you could come &
> make us a visit—. . . you shall have a seat either in my little "study," or
> in our larger parlors, made cheerful every fair day by the light of the sun
> from morning to night, with an uninterrupted view of the Charles wind-
> ing through the meadows, & hills of Brighton and Roxbury beyond, &
> in the distance a sight of the Blue Hills of Milton, near which we [Harris
> and his wife] passed our youth. This little cottage is a matter of wonder
> to the people here, from its simplicity, novelty, & cheapness, for it cost
> only three thousand dollars. . . . It is built, as New England homes gener-
> ally are, with a wooden frame, boarded & clapboards—but it is finished
> in the best manner, & has a solid granite foundation. Some call it the
> Quaker-house, or the broad-brim, on account of its gray color, its neat-
> ness, & the absence of superfluous ornament within & without, & from
> its projecting eaves, . . . We have two parlors 18 × 15 & 17 × 15 on the
> south side, connected by large sliding doors, so as to make one long
> room, with two fireplaces, & windows down to the floor, opening upon
> a light latticed veranda. The other rooms on the first floor are a dining
> room & my study, which communicates, immediately by doors with one
> of the parlors but also another door leading into the entrance hall. In
> the 2d story are 4 chambers with ample closets & 2 dressing rooms; & in
> the attic four very good bedrooms. The [illegible word] room & kitchen
> are in the basement, . . .[170]

Harris's daughters continued to live in the house until the late 1920s.[171] But by then the setting in which Harris took so much pride and delight was gone. Writing about 1908, his daughter reflected that "the small linden trees on the west, set out by our father, have grown into noble proportions. . . . [But] Of the view from the windows, nothing of the early beauty remains—the green marshes and wide meadows have long since been replaced by streets and houses—a swarming settlement hiding from us the winding Charles and opposite hills."[172]

The nature of Harris's entomological work in the era after 1842 will be considered in the next chapter. But some idea of the character and extent of his reputation in this period and how, in one case, it led to an inquiry about a relocation and new position, is shown in his relations with Thomas Affleck. Affleck's first letter to Harris from Ingleside, near Washington, Mississippi, was written in September 1846.[173] Affleck (1812–68), Scottish-born emigrant to the United States in 1832, is noted as an agricultural editor and writer, and he had one of the earliest southern commercial nurseries. He later moved to Texas.[174] Affleck asked Harris to assist him in investigating the cottonworm, which was devastating that vital crop in the South, proposing that Harris prepare the scientific description of the insect, based on specimens provided, while Affleck would contribute information on the activities of the insect and the crop damage.[175] It is apparent in this proposal that Affleck appreciated the technical skill and knowledge he and others in his part of the country lacked and which Harris had demonstrated in his publication, but he appreciated equally, perhaps more, the implications for agricultural improvement in a collaboration with Harris, commenting, "Your own work has done more to draw attention to [entomology] . . . than anything that has been written in this country. I am somewhat of an enthusiast in everything connected with the improvement of Agriculture, & have done much to stir up the South-West in that cause."[176]

In his first letter, Affleck also tried to entice Harris to point his interests and skills southward more generally, asking "Is there no possibility of your visiting & spending some time in the South, with the view of giving us something similar to your Mass. Report?" Affleck added that he would welcome any naturalist into his home, so long as the gesture would help to promote science.[177] A short time later, when Affleck encountered a published notice of Harris's plans for a revised edition of the *Treatise* on injurious insects, he repeated his invitation to visit, although his advice to "leave a substitute in your Library" shows a fundamental lack of knowledge of Harris's situation.[178] The two men continued their periodic exchange of letters, and in early 1850 Affleck wrote, "What sum would induce you to come here for, say, three years, as Professor of Entomology & Botany in Jefferson College? I do not say that an offer can be made you—but please answer the above."[179] Harris responded that the poor health of his mother and obligations to his wife and children would prevent his leaving Cambridge for any length of time, and added that, "It is true that I have to labor incessantly for my living, and have hard work besides to maintain my family on my college pay. But so long as health remains, and I am continued in office, I must continue to do my duty, as well as I can."[180]

In the meantime, another appointment was made in zoology at Harvard in 1847, but to this Harris's reaction was quite different than it had been to Gray's in 1842. Harris's English friend Edward Doubleday reacted with disgust at the appointment of Louis Agassiz, the Swiss-born scientist, to a professorship in the new Lawrence Scientific School at Harvard, seeing it as pandering to foreign scientists to the neglect of native Americans (presumably Harris himself).[181]

Harris, in fact, felt no anger or sense of injustice and seems to have been as taken with Agassiz as were so many others, referring to his "prepossessing manners" and noting that Agassiz had "done much to make science popular here, & has been a very active & efficient member of the Natural History Societies of Boston & of Harvard College, & of the American Academy." Harris made the point that he never expected an appointment in the Scientific School and had shown no interest in it. He further stated that he also knew "full well that my services were more important to the University in the office which I now fill."[182] The last is an important gauge of how his thinking had changed since 1842, for he now felt valued as librarian even though his entomological efforts were difficult.

Clearly, Harris had to adjust his scientific interests and activities to the reality of time and resources after 1842. In this later phase of his life, he seems to have had a less settled program in science and was more susceptible to external influences or to the momentum from the past. Still, science was engrained in him and in 1854 he wrote that, "Although my official duties, together with domestic cares & anxieties occupy most of my time, my interest in Entomology remains as strong as ever."[183]

MARRIAGE AND CHILDREN

As he notes, Harris's attempt to build a career, or to structure his life so that he could produce an adequate income and leisure time to carry out his entomological studies, also involved a commitment to his marriage and a growing number of children. Harris may have known Catherine Holbrook from a fairly young age, probably through the friendship of their fathers.[184] Catherine Holbrook was born in Milton in 1804[185] and attended school there and later went to boarding school at the Ladies Academy in Dorchester. She was talented at drawing and had lessons in watercolor painting in Boston at an early age, and was notable for her embroidery. Her daughter thought she had "decided talent. Indeed she had true artistic feeling and instinct; perhaps this is what gave such a perfection to all her handiwork, whether with needle, household work, or anything she did. Who ever tied a bow with more elegance, who ever arranged flowers with such grace! There was certainty and exquisite finish in every touch."[186] There is no available evidence that touches on the question of whether or not she ever used her artistic abilities to assist her husband with his entomological sketches.

Catherine Holbrook's upbringing was one of advantage. Dr. Holbrook had met the Marquis de Lafayette while serving as surgeon with the Continental Army, and when the French hero of the American Revolution visited the United States in 1824, he went twice to the Holbrook home in Milton and was introduced to Catherine as her father's baby. Later that same year, on November 15, the twenty-year-old girl and the twenty-nine-year-old doctor-naturalist were married in the parlor of the bride's family home, she described some years later as "a beautiful little bride" and he as "tall, slender, with wavy brown hair and dark blue eyes, . . . hardly less comely." Some fifty guests and caterer, plus servants, were in attendance.[187]

All his life, Harris had to concern himself with money, and the problem grew as twelve children were born between 1826 and 1849. It is possible that one of the attractions in the library post was the dependable income. But to meet his financial needs as librarian, he had to engage in special tasks for extra compensation. In 1836, he was working on the unbound pamphlets in the library to increase his income, but it brought little return, amounting to $31.45 at the rate of "60 cents per hour, the usual price paid to instructors for extra services." Although Harris's salary had been raised from $645 to $1,000 since he began his tenure as librarian in 1831, inflation had greatly devalued its purchasing power so that he actually was less well off than before. To illustrate, he gave a comparison of some "necessaries of life," as they were when he became librarian and in the year of his writing:[188]

	1831–32	1836
Flour*	$6	$9
Meats	4¢ to 8¢	8¢ to 17¢
Sugar	8¢	12^1/₂¢
Butter	15¢ to 17¢	25¢ to 28¢
Girl's wages per week	$1.25	$1.75

Cambridge rents, in the same time period, had risen 33 percent.

*Harris uses symbols or abbreviations for units of measure—flour in barrels (bbl.), and meats, sugar, and butter in pounds (he apparently writes ℔).

As listed, Harris had a servant girl in his home, and in fact they brought a "trusty servant" of Mrs. Harris when the family moved from Milton to Cambridge.[189] It appears that such domestic help was not unusual for families in his social class, given his inclusion of the item among the requirements of life.

Though the year 1842 brought personal loss with the death of both Thaddeus and Catherine Harris's fathers, they also inherited some money and, as discussed above, the money from Catherine's father Amos Holbrook made it possible for them to build a new home.[190] At the end of the 1840s, Harris still was required to carry out extra tasks in the library in order to meet his expenses, working, in this instance, on the supplementary catalogue. He appealed to the treasurer of the university for help with his personal situation, pointing out that, "One thousand dollars a year is <u>not quite enough</u> to maintain comfortably & decently a family of ten children upon, in so dear a place as Cambridge; . . . it is said that some of the gentlemen of the University, who have double that sum, do not find a superfluity of cash in their pockets at the years' end."[191] From this comment, it is apparent that Harris's salary had not changed from what it was in 1836.

Over the years, Harris was able to find some ways to make small supplements to his personal income through his scientific work. The most important were his lectures to the Harvard students. The other means of making money from science were minor compared to this, and overall no doubt cost him more than they profited. In 1839 Harris was engaged in a concentrated effort to put his entomological knowledge to profit. "I find it absolutely necessary for me to give up all other employments during my leisure hours this winter to write for compensation, & therefore shall have to defer publishing any descriptions or rather preparing them for publication at present. Money must be had, & I can only get it by writing." Of particular significance, Harris refers to the preparation of his state report on injurious insects as among the projects from which he hoped to achieve financial gain.[192] As one of the commissioners on the Zoological and Botanical Survey of Massachusetts, and in payment for the report, Harris received $350.[193] Despite all his efforts to maintain financial solvency, when Harris died his widow was left with the continuing problem of maintaining their home with some of the children still young and now fatherless.[194]

The Harrises' first child was born on January 25, 1826, and eleven other children (including twins) were subsequently born to the couple, the last in 1849, when Harris was nearing fifty-four years of age.[195] The oldest child, William Thaddeus, was described by his father in 1830 as "an unfortunate child, having had a disease of the intervertebral cartilages from birth," with permanent spinal curvature.[196] Although Harris was concerned for a time about his son's mental development, William graduated from Harvard College in 1846 and studied law. He died in 1854, at age twenty-eight, "after a painful illness of 8 weeks," but his father observed in consolation, "We shall miss him much; but his has been a happy exchange, for

which he had long been prepared, & to which he was fully re-
signed."[197]

Although he had referred to the birth of his second child as an
event that increased his "anxieties & cares,"[198] when she died unexpectedly before reaching her first year[199] Harris emphasized, perhaps in contrast to his son, that she was "a charming child, perfect
in form, active in motion, & healthy in constitution."[200] Harris's
other children outlived him, though his third child, Harriet, the
wife of astronomer and later director of the Harvard College Obser-
vatory, George Phillips Bond, died in 1858, just short of her thirtieth
birthday.[201]

Of Harris's six sons, only two followed their father and paternal
grandfather as graduates of Harvard College (William in 1846, as
noted, and Thomas in 1863).[202] Harris concerned himself about the
education of all his children, although he does not seem to have had
any great personal involvement. In 1834, he noted that during winter vacation he was teaching the three oldest since they did not have
a regular school at the time. However, it fell to Catherine Harris to
give most of the instruction.[203] Perhaps about 1846, Harris entered
into correspondence with an unknown person regarding the education of one of his daughters, pointing out that, although she had not
studied Latin, he had no objection if she undertook that language.
Apparently the intent was to see if the daughter might be prepared
to become a teacher, Harris commenting that "she has taste, but not
talents of a high order." In a sentence at the bottom of the draft of
the letter (but then crossed off) Harris admitted that "My occupations have always engrossed so much of my time that it has been impossible for me to give her any [regular?] instruction or much
assistance in her school studies."[204] In 1848 he reported on the
status of education, public and private, in Cambridge, for the benefit of his Mississippi correspondent Thomas Affleck. At the time, he
noted that one of his sons was attending the public high school,
while three others were enrolled in a private grammar school (at a
cost of $32 per student annually). A daughter was preparing to teach
(perhaps the daughter discussed above) and walked three miles to
Boston every day to attend the school of George Barrell Emerson
while costing her father $100 per year; she planned to complete her
studies in the normal school. Another daughter was attending a private school apparently in Cambridge. He noted further that in the
high school "are taught all the upper branches of an English education, & also Latin, Greek, & French, which may be learned by girls
as well as boys in this school. It is not uncommon to find girls here

who can read Latin as well as boys of their own age. All the girls in
Mr. Emerson's school are <u>obliged</u> to learn Latin & French."[205]

Harris opened his home to visitors and colleagues and his wife
and children participated in the hospitality. Apparently John Langdon Sibley, prior to his regular appointment as assistant librarian in
the University Library, was accustomed to staying or perhaps dining
with the Harrises.[206] Entomological colleagues also were welcomed
into the Harris home and came to know not only the parents. In
June 1841, Charles Zimmerman wrote from South Carolina, characterizing the children as he remembered them: "[H]ow is your family
. . . the sweet little pretty girls as well as the hearty boys: Harriet,
Emma, Kitty, merry Charles, strong 'what's his name?' [Clarendon?]
gentle Holbrook, sober minded Thaddeus, and the baby boy [Ed-
ward]."[207] A dozen years later, Zimmerman, in fond remembrance,
thanked Harris for his report on the "progress of your little ones to
manhood,"[208] a reflection also on the pride and pleasure that Harris
took in his children.

Of the guests in the Harris home, none seems to characterize the
concern and generosity of the couple more than the hospitality
shown the eccentric foreign-born bachelor, Evangelinus Apostolides
Sophocles, who became teacher of Greek at Harvard in 1842.[209]
Sophocles lived in Holworthy Hall and was found ill by Harris in his
room, apparently with typhoid fever. With his wife's concurrence, he
brought him home for nursing and recovery. Thereafter, Sophocles
became a part of the family, retaining his dormitory room while taking meals with the Harrises. His presence was said to have added a
zestful flavor to the Harris home life including argumentative conversation between the father and the Greek scholar over political
questions.[210]

Harris's daughter's account, written many years later, captures
something of the pattern and tone of his homelife:

> He passed busy evenings often, in "the study," writing into the small
> hours, or joined wife and children by the cheerful woodfire in the par-
> lor, when Harriet's guitar and his own flute and flageolet made sweet
> music—sweetest of all was his own baritone voice, in which he sang to
> his children hymns, or old ballads. One of the many happy remem-
> brances of my childhood is that of early evening hours, when, with his
> family about him round the hearth, or on the verandah, Father would
> devote himself to their pleasure. . . . he told his younger children—and
> he was a prince of story-tellers—tales of his own childhood. . . . it was
> part of our grandmother's code that the children of gentle-people
> should early be taught to dance, and our father, occasionally, with great

drollery, gave us a specimen of the fashion of dancing which prevailed in his childhood.[211]

From what little evidence is available, Harris and his wife appear to have had a close and supportive relationship. Early in their marriage, Harris reported reading from a novel to his wife,[212] while in the last year or so of his life his daughter notes that the couple often would sit hand-in-hand before the fire after dinner. Catherine, on her part, "strove by her own labor to save her husband's slender purse and be to him a true helpmate," in spite of the fact that she had been raised in an atmosphere where money was not a pressing concern. When Catherine died on the day before Christmas in 1887, she had passed nearly as many years in widowhood as she did as wife.[213]

Of Harris's daughters, reference was made above to Harriet's marriage to George Phillips Bond and her early death in 1858. Emma (1830–1930), Catherine (1832–1900), and Elizabeth (1844–1939) did not marry and all continued to reside in Cambridge. The youngest child and daughter, Sarah Harriet (1849–1935) married and in 1887 was living in Maryland. All of the sons, with the exception of William, married. In addition to graduating from Harvard College, William (1826–54) also received a law degree in 1848. In his brief life, he was editor of the *New England Historical and Genealogical Register* and assistant librarian in the Boston Athenaeum; in the summer of 1851 he also worked in the Harvard library.[214] Charles (1832–1909), the twin brother of Catherine, was a civil engineer who served as superintendent of streets in Boston, apparently after residing for a time in the West, where he may have learned his profession. Son Amos Holbrook (1834–1913) was a merchant in Boston. Clarendon (1836–92), also a civil engineer, moved to Chicago and in the 1880s was in Texas; he died in Cambridge. Edward (1839–1919), who lived for a time in Cambridge, later moved to Saratoga, Brooklyn, and Yonkers, where he died. As an architect with the firm of Ryder and Harris in Boston, Edward in 1864 prepared rough plans for a new library for Harvard University and consulted on the library building with his father's successor, John Langdon Sibley, in 1865 and 1870. The youngest son, Thomas Robinson (1842–1909), a graduate of Harvard College (class of 1863), served in the Civil War. He attended the General Theological Seminary in New York and entered the Episcopal priesthood; his ministry included service for a number of years in a church in the Bronx, New York.[215]

In 1849, Harris reported to his friend Edward Doubleday that his nine-year-old son, and the English entomologist's namesake, had discovered a larvae that "removes the insect to another group."[216] Edward Doubleday Harris, in fact, seems to have been the only one of Harris's children to show any adult interest in entomology, but his pursuit of insect studies did not occur until after he was sixty years old. An active member of the New York Entomological Society, he specialized in the Coleopterous family Cicindelidae (tiger beetles) and published a number of papers on the group. Edward died in 1919.[217]

Though Harris was given to long hours of work—at his medical practice, in the library, and on his entomological and other studies—his health was not always good and he complained in 1828 of headaches "which keep me always thin in flesh."[218] His son, in fact, reported that Harris was afflicted all his life with nervous headaches, but illness rarely confined him to home.[219]

3

Work in Science and in the Library

ENTOMOLOGY IN AMERICA

WHEN HARRIS ENTERED UPON HIS ENTOMOLOGICAL ENDEAVORS, THE SCI-entific study of insects was well developed in Europe, where naming, description, and classification of insects were the predominant interests.[1] The European enterprise encompassed American species, a fact that became an irritant to American students trying to establish an indigenous study of nature, and Harris felt the pangs of nationalistic competition as much as any naturalist of his generation. In the decades prior to Harris's initial work, there was some sporadic attention to entomological studies in America that appeared in the emerging scientific and agricultural periodical press. Virtually all of the (relatively few) publications from the latter part of the eighteenth century related to agricultural interests,[2] beginning with a paper by Virginia planter Landon Carter (1710–78) on the fly-weevil and means of controlling its wheat damage.[3] That early stage of entomological study in America was essentially individual, originating in the immediate interests of the authors rather than part of a concerted or organized approach to the subject.

Harris's mentor, William Dandridge Peck, began his work on insects in 1795 (the year of Harris's birth) with a paper on the cankerworms[4] that fed on leaves of fruit and other trees. Through the agency of Harris's subsequent work, Peck is seen as having instigated the serious study of economic entomology in the United States. In the early years of the nineteenth century, papers on entomological topics by various authors appeared in American publications, often relating to insects of agricultural or medical interest and generally to single or seemingly closely related species, again reflecting an individualized interest in insects and their relations to economic concerns.[5] Clergyman Friedrich Valentin Melsheimer (1749–1814) produced the first separate publication on insects printed in the United States. His *Catalogue of Insects of Pennsylvania* (Hanover, PA,

1806) related only to beetles but included more than 1,300 species with notes on their habits. His son, Frederick Ernst Melsheimer, one of Harris's correspondents, inherited his father's insect collection.[6] John Eatton LeConte also contributed to early American entomology and published a paper on the subject in 1824, where he pointed to the difficulty of pursuing the study of American insects while ignorant about what European entomologists already knew.[7]

Sustained interest in description and classification of insects in the United States began with Thomas Say, although he also followed other authors by occasionally writing on topics of economic interest, including the Hessian fly. Say was actively engaged in studying natural history, and his field work included service as zoologist with the expeditions of Major Stephen Long to the Rocky Mountains (1819–20) and the sources of the Minnesota River (1823); he was also a founder and curator of the Academy of Natural Sciences of Philadelphia. At the time of Harris's first acquaintance with Say, the older entomologist was involved in the preparation of his most ambitious work on native insects, the elegantly illustrated three-volume *American Entomology, or, Descriptions of the Insects of North America* (1824–28). When Say issued the first part of the work in 1817 (later reprinted with the 1824–28 publication), it was his ambition to include all known American insects.[8] Circumstances, including the cost and difficulty of production of the work, and Say's removal to the utopian community at New Harmony, Indiana, in 1825, prevented completion of whatever plans he might have had for continuation of the project as first conceived. Say's entomological work was devoted especially to the description of new species of insects (he named some 1,500) and demonstrated little interest in larger questions of relationship or classification. Likewise, he had little interest in life histories, habits, or control of insects (the approach so central to Peck and others on the agricultural side in the early study of insects in America) but historians have emphasized that Say's activities were a necessary foundation to the development of an effective agricultural science.[9] Harris agreed with the latter point.

At the outset of Harris's scientific career, it is clear that entomology was slowly emerging in the United States as a viable subject for investigation and was an area of activity with potential economic and nationalistic implications (as men like Say tried to compete with foreign naturalists in the description of American insects). Of course American nature was made up of a multitude of organismic forms and by most accounts insects were among the least studied. This situation presented both challenge and opportunity. Evidence of the relatively small amount of attention to entomology in America during the late eighteenth and early nineteenth century, when compared to the study of other subdivisions of biology, comes from a scan of the entries in Max Meisel's bibliography of natural history for the years 1768 to 1830. During that period about 1,800 articles, pamphlets, and books are listed, of which only about 100 were on insects.[10] In a review of American natural history in 1826, James Ellsworth DeKay concluded that zoology in the United States in general was less studied than botany, mineralogy, and geology. DeKay thought a manual for beginning entomologists was an important need, one that was only partially met by Say's *American Entomology*, since "its expensive form puts it beyond the reach of most private individuals."[11] As discussed below, Harris concurred in this assessment[12] and devoted time and effort to the promotion of interest in the study of insects.

If the conclusion by a contemporary entomologist is accurate, the relative position of entomology among the natural sciences did not change greatly during the two decades of Harris's greatest activity. In 1846, John G. Morris wrote bluntly that "entomology is of all branches of natural history the least cultivated by scientific gentlemen in our country," a situation that contrasted strongly with the popularity of insect studies in Europe.[13] The situation changed only slowly, but nonetheless steadily, during Harris's lifetime. Historian Conner Sorensen, for example, using "working entomologists" as a gauge, estimated the number at about twelve in the 1840s and thirty in the next decade.[14] As the field grew, the connection to agriculture became the most salient feature of American entomology as compared to Europe. By the 1870s, nearly half of the publishing, professional entomologists were sustained primarily by agricultural interests and the leadership in that group helped redirect entomology toward professional status.[15] According to Sorensen, however, the systematists and agriculturalists maintained the fabric of a scientific community until after the passage of the Hatch Act in 1887, which allocated federal funds to support state agricultural experiment stations.[16] How Harris dealt internally with the relations of taxonomic and economic entomology is an important aspect of his life story, but it would not have occurred to him that he should choose one over the other.

SCIENTIFIC WORK AND CONTRIBUTIONS: CA. 1820–35

Harris's serious scientific career began in 1823, when he published his first article and initiated a correspondence with Thomas

Say. In June of that year, he published "Upon the Natural History of the Salt-marsh Caterpillar" in the *Massachusetts Agricultural Repository*, and in July he wrote to Say at the suggestion of Thomas Nuttall. Both events give a good picture of Harris's outlook at the origin of his public career in entomology. His primary interests, as stated to Say, were focused on the life histories of insects "as appear to be injurious, or promise to be useful," rather than in collecting for the sake of curiosity alone. He described his interests as botanical and entomological and the insects with which he was concerned were noted for their damage to particular trees, including fruit, or of medical significance.[17]

When Harris first wrote to Say, he reported that only about ten percent of the species in his collection of five hundred were identified,[18] but five years later those identified were closer to a thousand.[19] He was making progress through hard and methodical work, developing record-keeping procedures that he recommended to his correspondents with whom he exchanged specimens and other information.[20] He wrote to Nicholas M. Hentz in 1829 to ask whether he set traps for insects. As Harris explained, the idea occurred to him when his New Hampshire friend and entomological supplier Levi Leonard mentioned procuring a number of insects that were stuck to the sap of maple trees. Harris suggested to Hentz that a "mixture of molasses and gum water in a shallow plate" could be used to attract small Coleoptera.[21] This general idea was one that arose in various places and which, overall, historian David Elliston Allen has called the "most dramatic tipping of the scales in favour of the hunter" before the mid-twentieth century. Edward Doubleday and Henry Doubleday, in England (the former of whom would became a close entomological friend of Harris), are credited with the idea of "sugaring" around 1832, which they used for the capture of moths, but the more developed and widespread practice of brushing tree trunks with a sweet concoction does not appear to have become common practice until the 1840s.[22] Whether or how Harris actually used the method is not determined; in light of his special interest in the nocturnal Lepidoptera, it would be interesting to know whether it ever occurred to him to use the method for capture of those insects, which apparently was the chief target group among the British entomologists.[23]

In a package of insects sent to Say in 1824 for identification or naming, Harris's sense of the relations of the insects to their habitat or food source was manifest. In regard to two insects from the plum and juniper tree, he instructed Say that "if they are not already described by some other name, [they] should, according to a rule of

Linnaeus, bear for their specific designations the generic names of the trees which afford them sustenance. . . . It appears most proper to distinguish the species of insects by the names of the trees or plants on which they live, in every instance where it is practicable."[24] Harris must have felt very strongly about this practice to have so instructed his teacher in entomology.

In the period before 1836, Harris published some thirty-three papers, twenty-four of which were in agricultural or horticultural journals (primarily the *New England Farmer*).[25] In the mid-1820s, one of the descriptive and taxonomic projects that Harris worked on was a study of a genus of scarab beetle (order Coleoptera), published as "Description of Three Species of the Genus Cremastocheilus," *Journal of the Academy of Natural Sciences of Philadelphia* 5 (February 1827): 381–89, but this was not typical of the bulk of his publishing in this period.[26] Though his published output was chiefly in the agricultural press, the great bulk of his entomological correspondence, on the other hand, related to classification and other technical aspects of the subject, and references to the destructive habits of insects or even their life histories are not prominent.

In the years 1828–29, Harris published a series of seven papers called "Contributions to Entomology" in the *New England Farmer,* prepared at the request of the editor, Thomas Green Fessenden, and in response to the many requests that Harris received for entomological information. Describing the series to a correspondent at the outset, he explained that they will be on insect "habits, to which will be added descriptions of new or ill-defined species."[27] Clearly, he intended to use this outlet not only to inform the public about life histories of insects but also to set forth new information (or supposed new information) on the physical character and names of insects.[28] The point, in this context, therefore, is that Harris did not see publication in the agricultural press as entirely separated from an interest in a more scientific study of entomology (whether of life histories or of taxonomy).

In his letter to John Lowell in 1827, which laid out a program of needs and priorities for American natural history, Harris stressed the intimate connections between insects and plants and the desirability of studying their "correlative" relations. These studies encompassed "the oeconomy, metamorphoses, & habits of insects" and their relations to humans. To Harris this "philosophical view of insects is what is most wanted & has been most neglected." While he goes on to say that arrangement, description, and naming are necessary first requisites (and his general correspondence underscores how much effort he gave to taxonomic questions in their own right), he also pointed out that Thomas Say knew nothing of botany and wrote little about the habits of insects.[29] Harris, thus, reached for a distinctive place for himself in entomological studies and by emphasizing life histories took the philosophical route.

Although Harris's publications tell important things about his natural history activity in this period, his rationalization and plans for publications that never came to full fruition in many ways tell even more. In the mid-1820s, Harris realized from personal experience that the lack of literary resources would inhibit entomological study and began plans for a work that would include descriptions of all the insects in his collection. Say had described many of the insects that he had considered to be new, and consequently he contemplated a visit to Philadelphia to enlist Say's help with genera and species names,[30] but this trip never occurred. The developing plans for what he called the Faunula Insectorum Bostoniensis were soon divulged to Hentz, along with his chief motivation to aid in the promotion of entomological studies among interested individuals.[31]

In 1826, Harris tried to organize an effort for a cooperative work among himself, Hentz, and Charles Pickering, essentially a compilation of descriptions of American insects from already published works, as a way to meet the general lack of readily available publications on the subject and to promote its study. Hentz accepted the proposition,[32] but Pickering, who was in the process of relocating to Philadelphia, found himself greatly disappointed with the available entomological resources there.[33] It is not clear that Pickering ever specifically declined to take part in the project, but his reaction to an April 1826 letter from Harris was not encouraging. Harris wrote to promote the virtues of a general introduction to entomology, which he described "as not dishonourable to the writer."[34] Pickering responded that the preparation of a popular work would take more effort than anticipated, and he thought Harris's time would be "more advantageously and agreeably laid out in examining the unknown parts of our interesting Entomology."[35]

In 1828, Harris was hard at work on the local fauna, modeled, as he said, on Jacob Bigelow's 1814 *Florula Bostoniensis* "but with plates."[36] He described the scope of the projected Faunula Insectorum as including "the most common & curious insects" and could more appropriately be designated as relating to New England rather than the Boston area alone.[37] In Harris's discussions of this work, practical concerns regarding the destructive habits of insects are not addressed, but personal as well as community considerations are revealed in his expressed hope that publication would help "to supply the increasing & imperious wants of a young family, as well as to furnish the lovers of Entomology <u>here</u> with assistance in prosecuting the science."[38]

The work, however, never appeared.[39] Later, in explaining the course of the project to his friend David Humphreys Storer, Harris attributed its lack of progress to the want of books for consultation although, he noted, some of his insect descriptions were published in the *New England Farmer*. In addition to the descriptive parts of the intended book, however, he also had prepared an introduction, largely adapted, as he states, from the British naturalists William Kirby and William Spence's *Introduction to Entomology,* on the external anatomy and life histories of insects. He explained to Storer,

> The difficulties met with at length led me to think of some means of making Entomology popular, and I looked to the young as the proper subjects to begin with. With the hope that by exciting a taste among children for this branch of natural history the parents might become interested also, I have rewritten my introduction in plain & simple language, divested as much as possible of all hard words, & intend to add to it brief descriptions of some of our most common insects. This you may think is small business; but I hope it may at least be useful & entertaining to those for whom it is intended.

There is no evidence that this work for children ever appeared and, if it had, it is unlikely that he would have wanted it to be known, cautioning Storer, "Above all say not a word about the book of insects for children, for my name will not be prefixed to it & I would not be known to have employed my time on such puerilities."[40]

It was not by his own initiative, but through the externality of a government-sponsored program, that some degree of recognition was achieved and a concrete product was realized from more than a decade of study. The catalogues of insects that Harris prepared for the two editions of Edward Hitchcock's *Report on the Geology, Mineralogy, Botany, and Zoology of Massachusetts* (1833 and 1835), in a certain sense, rescued him from a well-intended but not well-directed scientific effort during the 1820s and early 1830s.[41] The published catalogue, however, was a classified list of insect names only, without the descriptions that Harris had intended for his Faunula although the number of species was much more extensive than he could have contemplated for the descriptive work. In addition to Harris's list of insects, Hitchcock wrote in the second report that "In preparing the following Catalogue of our mammiferous animals, I have been permitted to make free use of notes kindly furnished by Dr. T. W. Harris, a gentleman so well known as an accurate zoologist, that the value of this Catalogue would have been much enhanced, could he have been persuaded to make it entirely his own."[42]

Harris writes in the publication that (except for about a half dozen cases) the 2,350 insect species in the catalogue were all in his own collection,[43] which means that the state list was synonymous with his own. In projecting what he had accomplished in comparison to what was left to do, Harris used a formula that he would reuse elsewhere. Drawing on an estimate made by another naturalist, that for each plant species there are six species of insects, and using the figure of 1,200 flowering plant species in Massachusetts, he estimated that the state was home to some 7,000 insect species.[44] One major event that distinguished the second from the first edition of Harris's list was his report in the second that Thomas Say "has done me the honor to name a large number of my new species," as well as providing identifying information on genera (especially in the orders of Omoptera and Diptera).[45]

Though Harris's catalogue was an enumeration of names only, nonetheless it was, in his eyes, a significant step forward for American entomology. While apologetic for the deficiencies, he believed that the 1833 edition was "the first instance in which the public has been presented with anything like a complete list of the insects of any part of our country," as Friedrich V. Melsheimer's *Catalogue of Insects of Pennsylvania* (Hanover, PA, 1806) related to the Coleoptera alone.[46] Historians Harry B. Weiss and Grace M. Ziegler endorsed Harris's assessment of his effort as a milestone in American entomology,[47] and Harris's nineteenth-century biographer said of the catalogue that "In the condition of American science at that day, it was a work of inestimable value."[48] In spite of its inherent value and the fact that the project cost him "a great deal of labor," he received no payment other than a copy of the reports and a few copies of his catalogues.[49]

Scientific Work and Contributions: 1836–42

By the mid-1830s, Harris was, in many respects, well situated to pursue his entomological research. He had published his catalogue of insects of Massachusetts, and even if it was not all he might have hoped, the effort brought some order to his work. In this he had the help of the most important American entomologist of the time, Thomas Say, who had assisted with the identification and naming of many of his specimens. Say, however, died before his assistance to Harris could be fully realized. Nonetheless, when Harris received

Say's entomological collection from the Academy of Natural Sciences in 1836, though in distressful condition, it did give him an advantage over all other American entomologists. About the same time, he also had full use of the Hentz Collection that he had helped to procure for the Boston Society of Natural History. In spite of these resources, however, Harris knew he needed other aids, and in addressing this practical matter he had to face some fundamental questions about his own knowledge and the relations of American and European insects. Harris subscribed wholeheartedly to the idea that no undetermined or unnamed insects should be sent to Europe, and as early as 1830 he confessed that a weakness in his knowledge of genera of insects was a primary reason why he felt unable to enter into an exchange with foreign correspondents of Charles Pickering.[50] One of his goals became the procurement of types of genera[51] as a resource for identification of his specimens, and this led him to speculate on the relationship of American and European insects. In response to a request from the British entomologist John Obadiah Westwood for "types of genera," Harris stated that he assumed that most American insects were related to European genera and, of the few that were not, his collection was incomplete and lacked duplicates to send.[52] This situation led to his puzzlement as to how he could help Westwood with his request.[53] Harris repeated his attitude toward European insects on various occasions, consistently looking upon them not as valuable in their own right but as aids to the identification of American species.[54]

Harris pursued the question of the relationship between American and European insects in his correspondence in the mid-1830s with Joseph Dalton Hooker. The British botanist was at the time only eighteen years old, although it is not clear that Harris knew that; at any rate he took Hooker's queries seriously, coming as they did with the compliment of Hooker's awareness of his catalogue of Massachusetts insects, the opening of an exchange of insects, and a gift of books from his famous father, botanist William Jackson Hooker. The young Hooker wrote to Harris at the suggestion of Francis Boott, the American physician, botanist, and Harvard graduate living in England. Hooker stated that "From a long & valuable list of Insects found in Massachusetts, which I have often run through with jealous eyes, I find that Britain ranks far behind North America in number of Insects as well as in beauty of specimens." He could not recognize more than two or three of the insects in Harris's list as also found in Britain and thought those probably were introduced. On the other hand, Hooker observed that many of the mosses that were listed in the Massachusetts report were familiar to him as also found in Britain.[55] Hooker had been interested in the study of mosses as early as age five or six, as he later recalled, and during the period of his correspondence with Harris he was actively studying entomology.[56] In Harris's reply to Hooker, he stated that, while a "large portion of our insects are referrable to European genera, as might be expected from our similar climate & latitude," American species are entirely different. As for the mosses, Harris thought the nature of their seeds facilitated dispersal, which would explain why Hooker found so many of the British mosses also in Massachusetts. For most animals and plants, however, Harris noted that the impassable ocean separating Europe and America meant that species on the two continents must be entirely the same (through introduction) or entirely different. He promised to send Hooker a "Discourse" in which he had taken up the topic of geographical distribution.[57] Whether the work ever was sent has not been determined,[58] but Hooker was interested enough to write to Harris a year later to remind him of his promise to send it since nothing else was available on the topic.[59]

Harris's views on the relation of American to foreign species remained constant, writing to an American correspondent in 1843 that "By comparing our species with those of Europe, I have found that many are much alike, but yet I think specifically distinct; & I adopt the principle, in all cases where a species cannot easily be shewn to have been introduced from Europe, to consider it as specifically distinct, or as not having sprung originally from the same stock, however close the resemblance may seem to be."[60] It appears, however, that on at least one occasion, Harris's skepticism may have led him to incautious conclusions, and this in regard to a plant at just the time that he was reeling from the realization that the Harvard professorship in natural history had gone to a botanist. In June 1842, Harris sent specimens of the plant in question, and apparently included his manuscript, "On the Lycopsis Virginica of Bigelow," to James E. Teschemacher, a Boston merchant and naturalist. In the manuscript, Harris cast doubt on the taxonomic placement of the plant, and added the general observation that

> Naturalists of the present day are much inclined to refer many of our American plants and animals to European species; and it is to be feared that this is too often done upon insufficient grounds. If we trust to such authorities, without due examination ourselves, we may fall into or continue to transmit errors as palpable as those pointed out in the foregoing communication [i.e., in the earlier part of his manuscript]. Moreover, similarity in habit, or in those external resemblances, which we might naturally expect to find in nearly allied genera or species, ought not to

be considered enough to establish their identity, especially when they are met with in productions so widely separated as those of the eastern and western continents.

Teschemacher responded that he could see no reason to introduce any uncertainty about the plant, which he doubted was new. Shortly thereafter, following "repeated examinations," Harris acknowledged his mistake and detailed the reasons why he had previously missed seeing the plant's relevant structures.[61]

Harris's study and publication plans in the mid-1830s were ambitious but not unequivocal and were subject to external pressures and opportunities. It was Charles Pickering, according to Harris's own near-contemporary report, who suggested that he undertake a "synopsis of American insects" after Say's death. Harris requested that Pickering consult Say's collection on his behalf for the project, and this led to Pickering's arrangement with the Academy of Natural Sciences of Philadelphia to have the collection sent to Harris. At first, Harris's idea was to prepare a descriptive catalogue based on the list of insects from his own collection that he had published in the second edition of the Hitchcock report, "for I was determined not to undertake to describe any insects but those which I had before my own eyes."[62] Harris reported to English entomologist John O. Westwood in June 1836 that, upon the arrival of the Say collection, he intended to begin work on the synopsis, using his own collection, that of Say, and the collection at the Boston Society of Natural History[63] (including the insects of Nicholas M. Hentz). As explained above, the condition of the Say collection upon arrival dashed his hopes of getting any substantial help from that source,[64] but the new resources near at hand, problematic though they may have been, appear to have redirected Harris's attention beyond Massachusetts or New England to the study of American insects more generally.

Harris, in fact, seems consciously to have considered himself the heir of Say among American entomologists. In 1835, even before receiving Say's collection, he was engaged in the revision of some of Say's "Harmony papers," some of which had already appeared in the *Transactions of the American Philosophical Society*. Of this undertaking, Harris observed, "I do not know a better service which could be rendered to American Entomology, or a better way of keeping in remembrance the labors & merits of my friend,"[65] a statement of selflessness on the one hand, but mild self-promotion on the other.[66] Say had solicited Harris's assistance in the publication of some of his entomological work by the Boston Society of Natural History, and Harris oversaw the posthumous appearance of two lengthy papers by Say (on the Coleoptera and the Hymenoptera) in the *Boston Journal of Natural History*, volume 1 (1835–37).[67] Harris was prepared to do more for Say's reputation and the promotion of American entomology in the process. In late 1836, he contacted Say's patron, William Maclure, through Say's widow, about the republication of the deceased entomologist's papers.[68] Harris was willing to do the editing, add notes, and conduct the work through the press. Although Harris and Maclure are said to have corresponded on the subject, the project never came about, Say's biographer suggesting that the reasons were economic.[69] In late 1837, Harris reported the rumor that a collected edition of Say's descriptions was underway in Paris.[70] In spite of Harris's normally strong feelings about Americans publishing their own natural history, he later confessed that he had no regrets that a French translation of Say's work was undertaken, not having had any luck himself in getting Maclure to support the work.[71]

In this period, as Harris's orientation in entomology was developing from a local one to a more general interest in American insects, he also had to consider his relationship to work as a generalist and specialist, emphasis on taxonomy and life histories, and ultimately to pure and applied entomology. In his letter to Joseph D. Hooker in mid-1836, referred to earlier, he spoke as though he were now working as co-author with the deceased Say, while explaining that "My own attention has been chiefly (& necessarily for the want of books & collections) turned to the history, habits, & transformations of insects—& less until lately to the determination of genera & species."[72] In trying to coax his friend Charles Zimmerman to "send me everything which you can spare from your collection of North-American insects," he explained the overall ambition that he was following, stating, "It has long been my ardent desire to publish descriptions of our native insects, & with this view I have collected specimens in all orders, have paid a good deal of attention to the larvae & their transformations, as far as I could discover them, & have collected a great mass of materials in the form of notes, memoranda, & descriptions."[73]

Harris's expanded view and involvement also led to some uncomfortable encounters with other's ambitions. Thomas Nuttall had promised Harris's student in entomology, John Witt Randall (Harvard 1834), that he would send the insects he collected in Oregon to Cambridge where they could be examined and described and where Harris hoped to find some of Thomas Say's Rocky Mountain species that had been destroyed in Say's own collection.[74] By the fall of 1837, Harris had in his possession the insects collected in Oregon by John

Kirk Townsend (who accompanied Nuttall there in 1834). Townsend had presented his collection to the Academy of Natural Sciences of Philadelphia and it had been placed in Harris's care.[75] Subsequently, however, Harris was caught in a controversy over ownership of the Townsend insects. In March 1838, he was informed that Townsend had reclaimed ownership of the Oregon insects—although Harris had been told by Charles Pickering when they were sent to him that they had been "collected for the Academy"—and that Townsend intended to sell them in Europe. In fact Harris had in his care other insects belonging to the academy and was proceeding to incorporate Townsend's with the others, with the "intention to prepare a <u>descriptive</u> catalogue of the whole, & offer it, for publication in the Journal Acad. Nat. Sc., as <u>a just tribute to the zeal of Mr. Townsend!</u>" He urged the academy to persuade Townsend either to give or sell the collection and offered to pay for the duplicates (if sold to the academy) or to buy the entire collection if the academy would not.[76] The academy successfully subscribed funds for the purchase of the collection,[77] but several months later, Harris reported that Randall had purchased certain insects from Townsend (which had not gone to the academy) and refused to re-sell them to Harris in order that they could join the others in the academy, and, in fact, stated that he was going to publish them in the academy's journal.[78] Finally, in August 1840, Harris reported to the academy that Edwin Willcox of New York had purchased the Townsend collection from Townsend and he had been requested to hand the insects over to the new owner. Harris refused to do so without the permission and directive of the academy, which was received in a communication signed by the curators, although he demurred until he received a concurrence from corresponding secretary Samuel G. Morton, who had signed the original agreement with him.[79]

During these years, Harris was involved in a number of conceptual projects that were realized only in part. In early 1838, he reported his plans and activities to his friend Charles Pickering, who would soon leave on his assignment as chief zoologist to the United States Exploring Expedition to the Pacific under the command of Lieutenant Charles Wilkes. Harris said he was working on a descriptive catalogue of insects in his collection, beginning with the Lepidoptera, which the American Academy of Arts and Sciences had promised to publish in its *Memoirs* with illustrations. He also had underway a catalogue of Coleoptera but he was uncertain how to get it published.[80] The geographical range of these efforts is not entirely clear. In late 1839, Harris repeated his view that his serious published efforts should relate only to specimens in his own collection, "where at any time I can examine & compare them in the progress of my labors."[81] Although this approach would feature insects of New England, by the late 1830s his exchanges and relations with other entomologists undoubtedly had added to his core regional collection a number of species from other parts of the country.

As discussed above, from the 1820s Harris was working in various ways toward the production of a manual that would help beginning entomologists. He had made efforts to persuade Say to undertake such a work, "which would serve for American insects, as Pursh's Flora and Eaton's Manual did for plants, and he [Say] assured me that he was collecting materials for the purpose."[82] After Say's death and the publication of his catalogue of Massachusetts insects in the Hitchcock report, it seems that Harris's efforts in the 1830s took on aspects of both his own long-term project and the one that he had tried to persuade Say to undertake. In August 1838, he told J. E. LeConte that he was preparing a descriptive catalogue "with the characters of the genera prefixed" and hoped it would be useful to the young, but, as in all of his entomological work, he faced the lack of research resources.[83] A year later, he presented a somewhat different vision of his project, explaining that "it is my intention to go on publishing on the subject of our insects hereafter, no isolated descriptions on supposed new species, which are of little service to Science, but monographs of the native insects in my collection."[84] In fact he claimed not to indulge in the description of single species, departing from "this self-imposed restraint . . . [only] in describing insects injurious to vegetation, in connection with a detail of their habits & transformations, & with suggestions of appropriate remedies against their ravages."[85] Although Harris refers to his work on the Coleoptera (beetles) during this time period, it was the Lepidoptera that particularly engaged his attention.

Scientific Work and Contributions: 1836–42: Lepidopterous Insects

Harris's work on the Lepidoptera, and especially on the moths, reveals how his scientific work proceeded and is a key to understanding his scientific career. Although the culmination came in the late 1830s, the interest and effort in this order began in the 1820s. The origin of the particular interest that Harris took in the study of the Lepidoptera is not determined, but he must have been strongly aware of the order from the beginning and including their relations to agriculture. His first paper was on a Lepidoptera species, the saltmarsh caterpillar.[86] In 1826, he reported that his friend Charles Pickering had urged him to study moths using the "same principles in investigating the genera that I have already done with the Papilionidae" (butterflies).[87] By 1828, he had identified the nocturnal Lepidoptera as a special interest and was spending much of his time on them. Through his friend Nicholas M. Hentz, he was seeking assistance from Ernst Friedrich Germar (1786–1853), a European expert on these insects.[88] Later that year, he reported to Pickering that over the summer he had prepared a paper on the genera of Papilionidae, Bombycidae, and Arctiadae (the latter two being moth groups) for the American Academy of Arts and Sciences.[89] In 1830, Say recognized that on account of the relatively unworked nature of the group, the Lepidoptera awaited an American author in counterweight to the efforts of Europeans, and "from the study that I think you [Harris] have devoted to that Order, it unquestionably waits your labour." Say offered to provide lepidopterous specimens (both European and American) and to publish Harris's work in the *Disseminator of Useful Knowledge,* which was issued at New Harmony.[90] The lack of resources to pursue his studies of the Lepidoptera, however, would continue to plague Harris and always set limits to what he could accomplish, and it did in this case.[91] Furthermore, Harris never presented nor published the American Academy paper described above.[92]

In the summer of 1834, Thomas Say wrote and speculated on "who will undertake the Lepidoptera, to describe them?" He told Harris that Titian Peale was working on plates that illustrated the stages of the insects,[93] but he had not been successful in persuading Peale to write the descriptions. In a letter Say wrote shortly thereafter, which may have been his last to Harris, he said "I am very glad to learn the probability that you will take in hand the American Lepidoptera." He promised to send Indiana species of the order, apparently species from other locations, and volunteered to provide descriptions from books to which Harris did not have access but regretted that he had no manuscript descriptions of his own to share and thus could not provide names.[94]

The next year, Harris wrote to the British entomologist John O. Westwood and suggested an exchange, welcoming especially the "types of the modern genera of the nocturnal Lepidoptera, in preference to any other insects."[95] Though the books on the Lepidoptera were expensive and the best were in German, a language he did not read, by late 1836 he was studying German in order to tap that source.[96] The next year he wrote to Francis Boott and complained that the lack of time and resources had hampered his work, and he had suffered the consequence that many insects that he considered new, including many he "had traced from the larva to the perfect state" were "subsequently described by European entomologists." But his solace was in the fact that "the nocturnal Lepidoptera of this country remain almost untouched by them; & hence I am the more desirous immediately to arrange & describe the native species in my collection."[97]

Harris was trying to get an overview of various groups of insects, practicing his belief (as referred to above) that work on individual species was of little value to science. But he also was faced with the dilemma of being anticipated in his own discoveries. He told Westwood that he thought his small collection of native Lepidoptera might include several "novelties" but he "always [had] been deterred from making them known for fear of encumbering the science with synonyms of species which, perhaps, may have been described in works inaccessible to me."[98] He was not always so reluctant to go forward with publication in the face of doubt, however, and his words to Westwood were given in the context of establishing a rapport with the English entomologist that would aid his work. In undertaking his series of "Contributions to Entomology" in the *New England Farmer* in 1828, for example, he admitted the likelihood that some of his supposed new species had already been described and named in one of the "many rare, expensive, or inaccessible European works on Entomology, or in the very brief and unessential descriptions of others," a fact that he would be happy to acknowledge when it became known to him.[99] The need to observe standards operative in the international scientific community, on the one hand, and to protect and promote American interests, on the other, posed a dilemma for Harris and other American naturalists and the ways in which it was addressed in the context of naming and describing specimens is taken up in the next chapter.

The letter of November 16, 1837, to Francis Boott, quoted above, was an effort to find a connection in England from which he could get a reliably identified set of nocturnal Lepidoptera and enclosed a letter to George Samouelle who was known to have insects for sale. To Samouelle Harris suggested an exchange, he giving Coleoptera, the most commonly studied insect order, for genera of British or European moths that he could use in his project to identify and describe North American species.[100] Boott, in fact, put Harris in touch with entomologist John Curtis, who indicated a willingness to provide Harris with the insects he desired in order to pursue his studies of American moths. In replying to Curtis, Harris pointed out that no one in the United States had attempted to name and describe the nocturnal Lepidoptera, and only John Abbot had collected and studied them. He mentioned the few published works available to him on the subject and asked whether Curtis might be willing to identify moths which he would send to him, observing that "From the few European lepidoptera now before me I suppose that by far the greater part of the American moths can be referred to European genera well known to you." But Harris also had something new to contribute, noting that "In regard also to certain insects which, at present, I suppose should be made the types of new genera, because there do not seem to be any corresponding to them in Europe, I am not always sure what are their natural affinities, or to what European genera they are most nearly related," and indicated that he would be pleased with the opportunity to have them examined by a knowledgeable European entomologist.[101] Curtis sent some three hundred Lepidoptera, taken from a collection he had initially prepared for a different purpose. He was willing to assist with generic names of insects that Harris might send, but did not have the time to identify them to the species level.[102]

During this period, Harris also attempted to mobilize American entomological workers to aid in his Lepidoptera project. For example, in early 1838 he wrote to Charles Pickering, who was in New York awaiting the start of the Wilkes Expedition, asking him to examine the collection of Dr. Wilkins,[103] who was said to have a number of American Lepidoptera, and to determine how far he had gone in identifying the modern genera of the nocturnal Lepidoptera. If Wilkins was accomplished, Harris wished to open a correspondence with him.[104] He also had hopes of getting help from the venerable naturalist-artist John Abbot, who resided in Georgia, and asked Nicholas M. Hentz to collect for him in Alabama.[105] Hentz accepted the request and also offered to prepare illustrations.[106] Potentially more important for his enterprise were Harris's solicitations from John Eatton LeConte, who in 1833 had published a work on North American Lepidoptera, *Histoire Générale et Iconographie des Lépidoptères et des Chenilles de l'Amérique Septentrionale* (Paris), written with the French entomologist J. A. Boisduval and with illustrations by John Abbot. Additional volumes of the work were projected but never appeared. At one point, Harris expressed hope that LeConte would transfer the Lepidoptera drawings that Abbot had done for him "together with his notes & specimens."[107] In the summer of 1838, Harris wrote to LeConte to see if he could borrow his named Coleoptera and said he had another proposition regarding the Lepidoptera that he would present later.[108] It was not until late 1839 that Harris appears to have directly approached LeConte about help with the Lepidoptera (though it is not clear whether his proposal then was the one he had in mind the previous year). At any rate, by the time of this contact, Harris's work on the order had taken on a renewed incentive and urgency through his contacts with the English entomologist Edward Doubleday (to be discussed below). Pleading a lack of knowledge regarding the larvae, pupae, and habits of certain species which was hampering his effort, he made a proposal to LeConte: "As the work on the Lepidoptera of North America, which was begun by you in conjunction with Dr. Boisduval, has for a long time been suspended, and as Dr. Boisduval seems now to have a task of such magnitude on hand, that he will probably never go beyond the Diurnal Lepidoptera of North America conjointly with you, I hope that you may be inclined to transfer your aid to me, or to join with me, if you so please, in completing the task, so well begun by yourself & Dr. Boisduval."[109] In what must have been his reply to this request (although it did not come until July 1840), LeConte referred to an unexplained break with Boisduval who "will not write to me," and noted that he did not himself have a complete copy of the work that he and Boisduval had written. In consequence, he was unable to answer Harris's questions on three Lepidoptera. By this time LeConte had decided to devote himself to the Coleoptera,[110] and apparently Harris never was able to get substantial aid from this most notable American student of the native Lepidoptera.

As indicated above, Harris knew that among American entomologists he alone was doing serious sustained work on the Lepidoptera (especially the nocturnal groups) without becoming a narrow specialist in that order. In doing so he developed a broad-based approach to the problems of classification and engaged in (what were for him, at any rate) innovative techniques for the discernment of species and their relations. By 1826, Harris was examining the nervures (or veins) in the wings of butterflies as an aid in classification, a procedure that had been used by the English entomologist Moses Harris as early as 1767. Though not without precedent among entomologists,[111] Thaddeus Harris's investigations seem to have been carried out along independent lines. In February 1826, he wrote to Nicholas M. Hentz that he had been spending a great deal of time on the Papilionidae, and, "I have discovered excellent generic characters in the nervures of the wings."[112] In response, Hentz offered his congratulations to Harris for his "discovery for the classification of the Lepidoptera by the nervures of the wings, as Jurine has done for Hymenoptera."[113] In a letter to Say in 1829, Harris tried to persuade him to forego further work on his ongoing illustrated *American Entomology* in favor of a work that would have greater public appeal and recommended the inclusion of tables of the wings of Hymenoptera, Diptera, & Lepidoptera, offering to let Say have his own drawings of the nervures of several Lepidoptera genera.[114] At the end of the 1830s, when he was seriously engaged in the study and classification of the nocturnal Lepidoptera, Harris wrote to his friend Edward Doubleday that "I believe that I showed you a set of drawings of the nervures of Lepidoptera which I made 20 years ago. They contain most of the genera of our butterflies, about twenty Bombyces . . . and some Geometrae. These drawings have materially assisted me in locating the families & genera, & give me additional confidence in the arrangements which I have proposed."[115] Expressed in these terms, Harris was not making a general claim for the validity of the use of wing veins in classification; his words seem to indicate how he personally used these features in reaching his own conclusions regarding the insects he had at hand.

The use of the wing nervures, however, was only a part of Harris's encompassing approach to insect classification. His study and utilization of the various stages of an insect's life and habits, in this regard, went beyond his concern for understanding how their life histories related to the interests of farmers in preventing plant damage. As indicated above, interest in life histories was a focal point for Harris's scientific interests from the outset and, judging by the products of his work as reflected in his bibliography, that interest was frequently directed to practical ends. But his correspondence also reveals how intimately he considered these factors in his classificatory work. Historian David Knight refers to government publications done in the United States in the nineteenth century on the natural history of insects and their relations to practical interests (no doubt having Harris in mind, among others), but Knight asserts a "gulf to be crossed" between those working from such interests and the mindset that is applied to natural classification.[116] For Harris, however, the relation between these interests seems to have been constitutive of his scientific work, so that life histories naturally became a focal point when he considered relationships and classification. Insofar as possible, he kept aware of developments in England and wrote to his friend Nicholas M. Hentz in 1827 that "the great naturalist, MacLeay" in developing his Quinary system, a natural classification based on series of five organisms related by affinity, had "laid great stress on the primary forms of larvae."[117] Harris referred his friend to the third and fourth volumes of Kirby and Spence's *Introduction to Entomology,* the chief work on the subject in English at the time, for more on the subject, and it is possible that this was his own source of information on MacLeay.[118] This broad approach to his entomological interests continued throughout his career, writing in 1852 that "In studying insects, it has been a primary object with me always to connect their habits with their organization, if possible; to find out the reason for this or that variation of structure, & what special use it is meant for in their economy."[119]

Thus, Harris's approach to classification of the nocturnal Lepidoptera drew on various features of the insects, in 1827 expressing a growing optimism about the use of the nervures as a guide in differentiating genera, but in conjunction with other aspects. "The neuration will afford good auxiliary characters to those taken from the palpi & antennae; the best of all are those derived from a knowledge of the larvae [caterpillars] & their habits; but these in the present state of our knowledge are least known, & to a beginner the least attainable."[120] In pursuing his studies of the larvae, he benefited greatly from information provided by his New Hampshire friend Levi Leonard who was dedicated to the investigation of insect habits.[121] Still his progress was hampered by lack of knowledge in this line. By September 1828, he reported that he had determined a number of subgenera within the nocturnal Lepidoptera, but, while he had completed drawings of the nervures from specimens in his collection, he explained that he had to "delay completing my labors in this department till I can obtain a better knowledge of the larvae and their habits; for where a marked difference obtains in these I have found a correspondence in neuration; and on these principles united (together with other easy characters taken from the imago),[122] I should wish the genera which I may propose to be established."[123] In the late 1820s, the goal for this work on the Lepidoptera was clarification of the insect order for his abortive Faunula Insectorum Bostoniensis.[124] Reference to the classificatory project does not reappear conspicuously in Harris's correspondence until the mid-1830s, when work on the Lepidoptera was again taken up as a special focus of his research program.

The years from 1838 to the career debacle of 1842, when he failed to get the appointment as professor of natural history at Harvard, were filled with entomological projects, and that on the Lepidoptera was an important part of the work in this period. First mention of a projected publication on the Lepidoptera in the *Memoirs* of the American Academy of Arts and Sciences comes in a letter in January 1838 where he mentioned the promise of the Academy's publication committee to publish the catalogue, including the authorization of an expenditure of $300 to $400 for preparation of colored illustrations to accompany the work.[125] Harris estimated that the project would take him at least two years to complete and, due to the general neglect of this order by entomologists, he expected to add "an immense number of new species & even many new genera." He told Nicholas M. Hentz that "Modern writers on the Lepidoptera make much use of characters derived from the larvae & pupae, from their metamorphoses & habits; on which account I beg you to [im]prove such opportunities as may present to obtain the former & take note of the latter. Honorable mention shall be made of any contributions which you may be induced to communicate to me respecting the history, habits, & transformations [of?] our insects."[126] He later argued his conviction that no arrangement of the Lepidoptera "will be true to nature" unless the larvae and processes of change are taken into account.[127] For the conduct of this work (as noted in the previous chapter) Harris imported books from Europe worth nearly five hundred dollars.[128]

Harris's illustrated catalogue of the Lepidoptera never appeared. Part of the reason must have been the important event that occurred in the fall of 1838, perhaps the most significant of his entomological career. The visit by the English entomologists Edward Doubleday and Robert Foster in the fall of 1838, and Harris's continuing relations with Doubleday thereafter, was at once stimulant, nationalistic incentive, and material aid in pursuing his entomological work and in particular his study of the nocturnal Lepidoptera. Doubleday and Foster apparently were with Harris in October.[129] In early November, Harris wrote to an Ohio collector, Charles J. Ward, and gave an account of the visit by the "two English gentlemen, Mr. Doubleday & Mr. Foster, of the 'Entomological Club of London'," who had spent a year and a half traveling in the United States and collecting insects, in Ohio but especially in New York and Florida. In particular, he referred to their collection of "<u>an immense number of Lepidoptera</u>," explaining that

> [T]he moths were mostly nondescripts, & some of them surpassingly elegant. Before leaving America these gentlemen spent a fortnight in examining and comparing insects with our collections here, & took note of all the species which they had collected & which they did not find in our cabinets, with the intention of describing their own new discoveries on their return. I made considerable exchanges with them, but not to so great an extent as I should have done had my duplicates been more numerous. You can judge of my mortification on finding that they would carry such a host of new things to Europe to be described there. This circumstance makes me only the more solicitous to go on with my own Entomological labors[130] and I hope during the next summer if not before to receive from you some of the new species in your region, which

may perhaps give me a chance to secure the priority in regard to some of the species.[131]

What Harris first thought of as a threat to American entomology and to his own project, in fact turned out to be an event that worked to his decided advantage. Doubleday encouraged Harris's quick publication of his work on the Lepidoptera in order to procure priority for his names. Conceding that Boisduval had preempted the field for butterflies, Harris began with resolve to produce a catalogue of the Sphinges (or Hawk-Moths).[132] By mid-January 1839, three months after Doubleday's visit, Harris had the catalogue nearly ready for publication, reporting to his entomological friend Edward C. Herrick that he intended to have it printed at his own "expense & risk"[133] Herrick, later librarian of Yale University, was working at the time with Benjamin Silliman on the *American Journal of Science,* and within a month it was concluded that Harris's paper would be published in that journal.[134]

Historian Harry B. Weiss noted that Harris published few descriptive papers, listing the one on the Sphinges and one other that appeared in 1853.[135] The Sphinges paper was published as "Descriptive Catalogue of the North American Insects belonging to the Linnaean Genus Sphinx in the Cabinet of Thaddeus William Harris," *Journal of American Science* 36, no. 2 (July 1839): 282–320. In the paper, Harris gives an overview of the character of the tribes, families, and North American genera of *Sphinges,* and in explaining his deviations from the practices of contemporary entomologists, notes that "After repeated trials [at arrangement], I have concluded still to adhere to the views of our great masters in Entomology, Linnaeus and Fabricius, especially as modern entomologists are by no means agreed upon the limits of the larger divisions of the Lepidoptera, and the order of the genera."[136] It appears, therefore, that Harris took a conservative and practical approach to his arrangement of the insects. Although the characteristics of the larvae, the insect's mode of transformation, and habits are an integral part of his description and differentiation of groups or species, the wing neuration is not. As Harris explained to a correspondent at the time of publication, the paper described fifty-five species, twenty-one of which he considered to be new, together with three new genera.[137] In the catalogue, all of the families bear Harris's names.[138]

The positive reaction to the paper by Rev. John G. Morris, who would later earn a reputation for his work on the Lepidoptera,[139] encouraged Harris's continuation of work on the order.[140] More important, however, was the encouragement and material aid from Edward Doubleday. Harris reviewed his relations with Doubleday in the fall of 1839. The English entomologist had "urged" him to publish his descriptions of Lepidoptera, "and said that if I would do it he would not allow the species which he had collected here to be described in England before the publication of my descriptions, but that if I declined the task, he should then with his friends of the Entomological Club of London, immediately set about describing our American Crepuscular [twilight] & Nocturnal Lepidoptera. Under the circumstances, I began with the Sphinges . . . and am now at work upon the Bombyces &c. Mr. Doubleday has sent to me a suit of all of them which he collected in this country, & also procured for me a set of drawings made by Abbot . . . many not in my collection."[141] Doubleday had found a small folio volume of eighty-four drawings of Georgia Coleoptera and Lepidoptera by John Abbot in a bookshop (apparently in London) and sent them to Harris. In broaching the probability of sending his insects to Harris, Doubleday had commented, as a vote of confidence, that "I had rather your papers were as complete as possible, and it would be far more advantageous to science that you should do them. You, too, can go on better, from greater knowledge of larvae, etc."[142] In turn, Harris admitted that Doubleday's help, advice, and encouragement were crucial to his work on the nocturnal Lepidoptera.[143]

By late 1839, therefore, Harris was well positioned to prosecute his studies of the nocturnal Lepidoptera. His plan was to take up the Bombyces (a group of moths referred to as spinners) and to treat them in a way similar to the Sphinges,[144] and then to undertake the Noctuae (owlet moths).[145] He was not simply engaged in the description of species but undertook to develop a schematics of relationships, as he informed Doubleday. "After much study, examination, & repeated trials, I have, without being fully satisfied myself, at length hit upon the following arrangement, which is confessedly imperfect, in as much as it is impossible to preserve, in a linear series, those varied approximations of forms, miscalled affinities,[146] which can only truly be represented by placing the groups, like towns on a map, & connecting them by radiating lines, like roads." Harris proposed the possibility of two series of the Bombyces, but "You will not suppose that I think these two series should follow each other, but rather that they are parallel, and both begin and end at the same point, running together or coalescing at their extremities,"[147] language somewhat reminiscent of the system of W. S. MacLeay (see above) as summarized by Mary Winsor.[148] He took up the subject a year later in another letter to Doubleday, giving more detail and presenting the relationships visually in various ways, including a circular arrangement that differentiated relations of affinity and analogy.[149] He repeated the scheme for John G. Morris shortly thereafter, pointing out that "This arrangement is based upon the forms of the larvae, their transformations, & the character of the winged insects."[150] In response to the presentation, Doubleday pleaded that "I can hardly find time to follow you through all the details of your proposed arrangement of Bombyces," but added some specific comments.[151]

In the late 1830s and early 1840s, Harris was at his height as an entomologist. It is of critical importance, therefore, to understand how his work on the Lepidoptera related to his overall research program, to events in his life (namely the failure to achieve the Harvard professorship of natural history), and his long-term historical reputation. Clearly, Harris is not known as a taxonomic entomologist and yet that seems to be the direction of movement in his Lepidoptera work during this period. But the emergence of Harris as the model practical (or economic) entomologist, latent in his work from the outset, played out in various ways, both subtle and obvious.

In 1840, Harris had other work on the Lepidoptera underway, namely for his report on injurious insects for the Massachusetts state commission, explaining to one correspondent that the report "will contain a good deal about our moths."[152] Harris's work on this report must have affected his attitude toward his studies of these insects. For example, in a September 1840 letter to Edward Doubleday, with whom his entomological exchanges generally were about taxonomy, he included a detailed and first hand observational report of a caterpillar's actions in relation to the cocoon within which it resided and Doubleday published it, with illustrations, in Newman's *Entomologist*.[153] Symbolically, at least, the letter to Doubleday can be seen as part of the transition that Harris made during the period, from the ambitious plans for his catalogue of Lepidoptera, to a time when he was deflected from that undertaking, to the elevation of his report on injurious insects as the defining outcome of his life.

It is clear, however, that factors other than personal career loss were at play in the failure to follow through on his Lepidoptera program. As always, lack of resources was an important consideration. In early 1842, just before events leading to the appointment of Asa Gray to the Harvard professorship began to unfold, Harris wrote to Frederick E. Melsheimer, "It was my hope, before this time, to have published a descriptive catalogue of our Bombyces (from Lithosiadae to Notodontadae inclusive), on the same plan as that of the Sphinges, but with outline drawings of some of the larvae. But I am yet deficient in my species, & have so much to learn respecting the larvae, that I have become almost discouraged."[154] Melsheimer, in reply, pleaded, "My dear Sir, we all look to you for classifying & naming our Lepidoptera—you only are qualified for this task. I give you my word that I shall do all which my feeble and limited power enables me to do, to aid you in your noble labors."[155]

In spite of the difficulties posed by insufficient information resources, clearly Harris, at the time, considered his work on injurious insects as only an interruption in his on-going scientific work on the Lepidoptera. His intentions were clearly stated to Doubleday in April 1840, where he vowed that "As soon [as] my 'Report' is off my hands I shall return to the Lepidoptera & you will then hear more from me on that subject."[156] Toward the end of that year, he wrote to Frederick E. Melsheimer and reiterated his intention: "As soon as my Report is printed I intend to resume my descriptive catalogue of moths, having already worked a good deal upon it, particularly on the Bombyces & PseudoBombyces (Arctias & Notodontes, &), & if my materials respecting the larvae were more complete, I should soon be able to print this portion of the catalogue. It is my present intention to add outlines of some of the larvae to the catalogue, & I have on hand a large number of drawings which I have made in years past."[157]

In late 1842, he wrote that "if I can finish a descriptive catalogue of our Lepidoptera, it is all that I can hope to accomplish."[158] But Harris never returned to the project with any commitment, and it appears that he did no serious or sustained work on the order (and in particular, on the nocturnal Lepidoptera) after the early 1840s beyond what was undertaken in revisions of his *Treatise* (as Harris's private reprinting of the state report was known). Nonetheless, he continued to claim a special place among American entomologists in relation to the order even after his active work was derailed by life circumstances and the perpetual lack of resources:

The study of the groups, tribes, & families, & genera of the Nocturnal Lepidoptera, & particularly of the Noctuans, is allowed to be very difficult, even by European Entomologists who have the advantage of large collections of species named & classed by the first naturalists, & numerous costly works for reference,—neither of which can a poor New England naturalist hope to see, much less to possess; but might be contented to labor onward, with such meagre helps as his own limited resources will allow him to command. No person in this country has undertaken to study scientifically our nocturnal Lepidoptera, & I am entirely alone in the pursuit.[159]

SCIENTIFIC WORK AND CONTRIBUTIONS: 1836–1842:
NATURAL HISTORY SURVEY OF NEW YORK

Harris's success as an entomologist by the late 1830s meant both opportunities and diversions. One not previously noted in biographical accounts is the work he committed to do for the Natural History Survey of New York. What is known is the origins of the project and negotiations about what would be acceptable. James Ellsworth DeKay, who was in charge of the zoological department of the survey, wrote to Harris in January 1838 proposing that he undertake the work on New York insects. The general outline of what DeKay had in mind was given: to include characteristics of families and genera accompanied by bibliographic references, and typical species of each genus with "detailed description accompanied by the specimen itself. In the choice of species preference to be given to one of your own or to a species described but not figured by Say."[160]

Harris made a rapid reply to DeKay's proposition indicating that he was "sincerely desirous of accepting it, not only on account of my love of science, but, I will frankly confess, because I am poor & need the compensation." He outlined his working methods and standards, explaining that "I never allow myself to give a list of names for insects submitted to me for examination until after I have most carefully compared them with the specimens in my own cabinet. . . . I hope in this way to do something toward producing uniformity in the nomenclature & synonyms." Because of the restrictions on his own collecting, and "demands made upon me in various quarters" for duplicates, Harris would not be able to send specimens to accompany the descriptions, as DeKay had requested, unless DeKay's collectors could provide those that he lacked. Harris agreed to provide the information as proposed by DeKay if "the little difficulties herein mentioned can be obviated."[161]

DeKay responded to Harris's questions before the end of February,[162] and by mid-March Harris informed a correspondent about the agreement: "I have engaged to furnish the characters of the Families & genera & descriptions of one species of each genus found in N. York & neighboring states for the New York Zoological report, which will be published in about three years from hence. Every species to be accompanied by a figure."[163] Harris made it clear, in corresponding with Edwin Willcox in New York, that the agreement with DeKay did not include the habits of the insects,[164] a particularly interesting point in light of the work Harris was doing about this time on the insects of Massachusetts. From information provided by Willcox, however, it seems that DeKay had recruited widely for assistance with the New York insects, and that Harris was not alone on the project.[165] In response, Harris insisted that his contribution was to provide descriptions of insects and he expected this to be distinctive from any assistance DeKay might receive from others in "making up his catalogues."[166] It appears that Harris sent a sample manuscript of his projected contribution to the New York Survey (by way of Willcox) and DeKay reported back that it "meets my entire approbation."[167]

The last letter between Harris and DeKay that has been found was sent in January 1839 in regard to DeKay's shipment of insects.[168] This was a year after the topic was first discussed between them. At the end of 1840, Harris still had the project in mind, but he was unable to undertake any concerted effort until his Massachusetts report was completed.[169] DeKay's zoological report appeared as part of the final reports of the New York State Survey, issued in five volumes between 1842 and 1844 and covered mammals, birds, reptiles and amphibia, fishes, mollusks, and crustacea, but not insects.[170] Harris's American successor in applied entomology, Asa Fitch, had applied to DeKay in 1839 seeking to work on the state's insects, but DeKay never responded, having already made the arrangement with Harris. Fitch's biographer writes that DeKay decided not to include insects in his report, but gives no reason. In 1841, he became ill and was thus afflicted for the remainder of his life, and in 1842, the New York governor largely terminated work on the natural history survey.[171]

SCIENTIFIC WORK AND CONTRIBUTIONS: 1836–1842:
MASSACHUSETTS SURVEY AND THE REPORT (TREATISE)
ON INJURIOUS INSECTS

Harris's various entomological projects during the late 1830s and early 1840s often did not come to full fruition. The one on which he fully succeeded was his *Report on the Insects of Massachusetts, Injurious to Vegetation. Published agreeably to an order of the Legislature, by the Commissioners on the Zoological and Botanical Survey of the State* (Cambridge, 1841). Here again, as with the earlier insect catalogue published in Edward Hitchcock's state-sponsored report, the support and incentive of a state survey resulted in a concrete product from his entomological labors that he did not otherwise achieve on his own.

The natural history survey of Massachusetts for which Harris prepared the insect report was, in certain respects, a continuation of the earlier geological and natural history survey to which he had contributed in the early 1830s.[172] The new survey was authorized by the legislature on April 12, 1837. A. A. Gould, in his historical review of the Boston Society of Natural History in 1842, gave primary credit to the society's council for the origin and conduct of the zoological and botanical survey which was connected with Hitchcock's geological survey.[173] George B. Emerson, who was president of the society and head of the commissioners, was advised by Governor Edward Everett to divide the responsibilities among the members according to their expertise.[174] The work of the commission was distributed as follows: David H. Storer (fishes and reptiles), William B. O. Peabody (birds), Ebenezer Emmons (quadrupeds), Chester Dewey (botany), Harris (insects), Augustus A. Gould (invertebrates other than insects), George B. Emerson (trees and shrubs).[175]

Overall, Harris considered the Massachusetts report, in conjunction with the work he was later asked to do for the New York state survey and the offer from the American Academy of Arts and Sciences to publish an illustrated catalogue of Lepidoptera, as an opportunity "of publishing descriptions of our insects without involving myself in ruinous expense."[176] In writing his report, however, Harris took seriously the governor's instructions to the commissioners for the survey to keep utmost in mind the benefits to agriculture. The governor advised that "it is not intended that scientific order, method, or comprehension should be departed from. [But] At the same time, that which is practically useful will receive a proportionally greater share of attention, than that which is merely curious."[177]

When Harris began his career in the early nineteenth century, the population of Massachusetts was overwhelmingly agricultural,[178] although then, and increasingly in later decades, commercial and industrial interests overshadowed the farming sector in the public sphere.[179] As the economy changed, the character of the agricultural sector also changed. During the early years of the century, residents in urban areas still maintained livestock and engaged in gardening, and the involvement of New England farms in commercial activity was limited.[180] The War of 1812 and the British blockade was a setback for sales of surplus farm goods as it was for New England commercial and shipping interests in general.[181] However, the growth of the population engaged in manufacturing and other nonagricultural activities in the years thereafter drew Massachusetts farm products into the market economy.[182]

In writing on the early years of the century, agricultural historian Percy Bidwell noted that "The New England region was by nature better fitted for grazing and pasturage than for agriculture in the strict sense of the word." Corn was basic to New England agriculture while gardening was little practiced by farmers at that time.[183] By 1840, however, "dairying, and the production of hay, corn, vegetables, and fruits were increasing" in the region.[184] So-called "market-gardeners," located near Boston and other cities, engaged in growing vegetables including the use of hotbeds that made it possible to produce more than one crop a year.[185] Apple orchards had been common to most farms where cider was widely consumed.[186] By 1859, Massachusetts ranked third of all the states of the Union in the value of its garden products and fourth among urban industrial states in the combined value of fruits and vegetables grown.[187]

The references to trees, vegetables, and other plants in the index to Harris's report on insects injurious to vegetation gives what may be a somewhat eccentric, but nonetheless useful, gauge of the botanical products that were of interest to agriculture and horticulture in the state in the early 1840s, as seen from the perspective of his study. A count and categorization shows the following distribution of those index references: trees, shrubs, vines (nonfruit), 33%; fruits, berries, 25%; vegetables (including cabbage, corn, cucumbers, onions, peas and beans, potatoes, squash, turnips), 20%; grain, grass, herbs, 14%; and flowers (garden and wild), 9%.[188]

From late in the previous century, attempts had been made to improve the quality of agriculture in the state through study and distribution of information, an effort that had mixed results and demonstrated class tensions in the Commonwealth. Although significant advances in knowledge and practice had been achieved in English agriculture, at the beginning of the nineteenth century this had had little effect on the situation in the New England region.[189] The Massachusetts Society for Promoting Agriculture, founded in 1792 in order to reform farming practices, was dominated from the outset by individuals more closely related to the commercial than to the farming class itself. Criticized and ignored by ordinary farmers who saw them as book-farmers detached from the day-to-day reality of agricultural activity, after about 1830, the agricultural societies turned away from their earlier efforts to reform agriculture in the state.[190]

Despite these tensions, advances in agricultural practice were increasingly evident, for example, in the improvement of the soil.[191] Harvard graduate, lawyer, and legislator Charles Theodore Russell,[192] though professedly not a practicing farmer, thought the major advance in Massachusetts agriculture in the first half of the nineteenth century was the "desire for improvement" among farmers.[193] What role books and journals played in this process is uncertain, and the attitude of the ordinary farm population toward the leaders of the state agricultural societies suggests that literary aids to progress were not easily accepted. But Russell sensed that there was a growing call for agricultural books, and even though works on scientific agriculture were not widely available, the agricultural journals helped to disseminate their message.[194] While the readership for the farm journals was far from universal, by the end of the antebellum period it has been estimated that one in ten American farmers were subscribers.[195] The founder and editor of the *New England Farmer*, Thomas Green Fessenden, promoted improvements in the raising of livestock and the growing of fruits and vegetables,[196] areas where regional agriculture made advancements in the early decades of the century. As noted above, Harris published a series of articles on entomology in the *New England Farmer*, and the journal published articles on insect damage by other contributors (though not in the systematic way in which Harris wrote). Reader concern with the fight against insects was one of the many topics that the *New England Farmer* addressed and clearly had a place in the economy of agricultural practice.[197]

Following the state's priorities in preparing his report for the Commissioners on the Zoological and Botanical Survey, and recognizing "the want of a work, combining scientific and practical details on the natural history of our noxious insects," Harris chose to feature the insects that were "remarkable for their size, for the peculiarity of their structure and habits, or for the extent of their ravages," a selection that in itself constituted "a formidable host."[198] As always, he turned to Nicholas M. Hentz for help, asking whether he would take charge of the spiders.[199] In explaining what he would expect from Hentz, Harris also gave a good idea of how he conceptualized the report at this early stage.

> What I want is <u>descriptions</u> of all the species which you have observed in Massachusetts, with their <u>varieties</u>—accompanied, either under the genera or species, with an account of the habits, & any interesting remarks which you can make relative to the economy; & also—a notice of what species are particularly venomous, with the remedies—& an answer to the question whether any species are now, have been, or can be usefully employed in the arts.[200]

The first part of Harris's report was done by spring 1838 and published along with those of the other commissioners.[201] Harris's contribution, some forty-eight pages, was the longest in the volume but dealt only with the Coleoptera.[202] It was issued as a house document and not offered for sale. Harris got about a dozen copies for his private distribution.[203] The survey was reauthorized for additional work in 1838.[204] By early 1840, Harris was working hard on his final report and had hopes of finishing it soon,[205] but completion of the project took longer than he anticipated. About this time, also, a political change took place and public concern arose regarding the costs of the geological and natural history surveys, which were referred to in the press as "lavish expenditures."[206] In February 1840, the legislature drew up a plan for general reduction of government spending and noted that the total of $5,500 authorized in 1837 and 1838 for a mineralogical, botanical, and zoological survey, "which it would seem might require a generation to complete," already was exceeded and only fishes, birds, and reptiles had been reported on. The legislative report recommended that no further appropriations for the survey be authorized.[207]

On February 15, 1841, the governor approved a resolution that effectively ended the survey as of May 25 of that year.[208] A July 1841 deadline was set for completion of Harris's report, and in April some 240 pages had been printed, but still he had 150 pages more to write.[209] In the meantime, Harris had concluded that since the state reports "are not like books published by private individuals," but are intended for distribution among the legislators, he would follow the advice of friends and reprint the report, assuming that Emerson as chair of the commission or the governor did not object. He intended to use the type set for the state report for his reprint and referred to reports printed by other commissioners in the *Journal* of the Boston Society of Natural History as precedent.[210] In the reprinting, he would include a new title page and running title "& shall make some slight changes in the body of the work—all which, with the cost of paper & press work, will of course be done at my own expense."[211] Following publication of the *Report,* which by title was limited to Massachusetts, Harris issued his own *A Treatise on Some of the Insects of New England, Which Are Injurious to Vegetation* (Cambridge: John Owen, 1842). Two hundred and fifty copies of the 459-page *Treatise* were printed.[212] Harris thought he would initially sell about fifty copies at not more than two dollars each, "& shall be content if these are disposed of, to let the remainder rest in my own hands till they are called for, & if no demand is made for them I can give them away . . . or sell them to the trunk-makers."[213] For various reasons (which he does not specify), but in order to avoid interference in the distribution of the state *Report,* publication of the *Treatise* was delayed until the fall of 1842. Already, however, he was contemplating the possibility of a second edition if the first sold well. He suggested to his correspondent Edward Herrick, on whom he had largely relied for firsthand information on the Hessian fly (an insect destructive to wheat, a crop not grown in Harris's region), that if a second edition were undertaken he would welcome additional help.[214]

The state legislature authorized a new edition of the report in 1850, to which Harris would hold copyright,[215] and it appeared in 1852.[216] Charles Flint, secretary of the Massachusetts State Board of Agriculture, was directed by the legislature in 1859 (three years after Harris's death) to issue a new and illustrated edition, and in 1861 he was authorized to use the plates prepared for that edition "in the publication of one or more editions designed for a wider circulation than that for the State could be expected to have."[217] It is this posthumous edition of 1862, from which geographical limits had been eliminated from the title, that is the basis of Harris's historic reputation, a fact undoubtedly aided significantly by the illustrations that were entirely absent from the editions issued during his lifetime. In the nineteenth century, several printings of the third edition of the *Treatise* were issued, especially by the New York agricultural publishing concern, Orange Judd and Company.[218]

Harris began his report[219] with the argument that knowledge of insects will go far toward developing means to control their damages, but the approach must be scientific: "Information on this subject is to be obtained by observation alone; it can be communicated and rendered useful to others only by means of correct descriptions of the insects themselves, accompanied by full accounts of their habits in every stage of their existence."[220] Appealing to historic authority, Harris pointed out that Linnaeus himself, "who, while giving to natural science its language and its laws, neglected no opportunity to point out its economical advantages."[221] At the end of the introduction, Harris takes up a defense of the study of natural history and in particular the place of scientific naming, arguing that "every intelligent farmer is capable of becoming a good observer, and of making valuable discoveries in natural history; but if he be ignorant of the proper names of the objects examined, or if he gives them names, which previously have been applied by other persons to entirely different objects, he will fail to make the result of his observations intelligible and useful to the community."[222]

The body of the work is organized by insect orders, and within each chapter he proceeds by scientific group (Latin genus and species names are used throughout, along with common English language names when available), presented, as he explained to a

correspondent in regard to the moths, "in the narrative style."[223]
The chapter on the Lepidoptera is the longest in the book, a situation derived from the fact that "there are perhaps no insects which are so commonly and so universally destructive as caterpillars,"[224] though also no doubt influenced by Harris's special interest in the order.

A short extract from the work is the best evidence of its content and style. Harris was working on the Bombyces (spinners) about the time he was finishing his report, and an extract from the writing on that group is of interest. His presentation is inevitably hierarchical, introducing, over the space of some pages, the Lepidoptera, then the moths, then the Bombyces, then (among others) the family Arctiadae or Arctians (known to the English as tiger-moths or ermine moths, most of whose caterpillars are called wooly bears) and the genus *Arctia*.[225] Finally, there is Harris's account of the species *Arctia Virginica*, including description of both the caterpillar and the adult moth:[226]

Of all the hairy caterpillars frequenting our gardens, there are none so common and troublesome as that which I have called the yellow bear. Like most of its genus it is a very general feeder, devouring almost all kinds of herbaceous plants, with equal relish, from the broad-leaved plantain at the door-side, the peas, beans, and even the flowers of the garden, and the corn and coarse grasses of the fields, to the leaves of the vine, the currant, and the gooseberry, which it does not refuse when pressed by hunger. This kind of caterpillar varies very much in its colors; it is perhaps most often of a pale yellow or straw color, with a black line along each side of the body, and a transverse line of the same color between each of the segments or rings, and it is covered with long pale yellow hairs. Others are often seen of a rusty or brownish yellow color, with the same black lines on the sides and between the rings, and they are clothed with foxy red or light brown hairs. The head and ends of the feet are ocre-yellow, and the underside of the body is blackish in all the varieties. They are to be found of different ages and sizes from the first of June till October. When fully grown they are about two inches long, and then creep into some convenient place of shelter, make their cocoons, in which they remain in the chrysalis state during the winter, and are changed to moths in the months of May or June following. Some of the first broods of these caterpillars appear to come to their growth early in summer, and are transformed to moths by the end of July or the beginning of August, at which time I have repeatedly taken them in the winged state; but the greater part pass through their last change in June. The moth is familiarly known by the name of the white miller, and is often seen about houses. Its scientific name is *Arctia Virginica*, and, as it nearly resembles the insects commonly called ermine-moths in England, we may give to it the name of the Virginia ermine-moth. It is white, with a black point on the middle of the fore-wings, and two black dots on the hind-wings, one on the middle and the other near the posterior angle, much more distinct on the under than on the upper side; there is a row of black dots on the top of the back, another on each side, and between these a longitudinal deep yellow stripe; the hips and thighs of the fore-legs are also ochre-yellow. It expands from one inch and a half to two inches. Having been much troubled with the voracious yellow bears in the little patch, (I cannot call it a garden,) where a few beans, and other vegetables, together with some flowers, were cultivated, I required my children to pick off the caterpillars from day to day and crush them, and taught them not to spare "the pretty white millers," which they frequently found on the fences, or on the plants, laying their golden yellow eggs, telling them that, with every female which they should kill, the eggs, from which hundreds of yellow bears would have hatched, would be destroyed. In some parts of France, and in Belgium, the people are required by law to *écheniller,* or uncaterpillar, their gardens and orchards, and are punished by fine if they neglect the duty. Although we have not yet become so prudent and public spirited as to enact similar regulations, we might find it for our advantage to offer a bounty for the destruction of caterpillars; and though we should pay for them by the quart, as we do for berries, we should be gainers in the end; while the children, whose idle hours were occupied in the picking of them, would find this a profitable employment.[227]

Harris's report was favorably noticed in the general and scientific press.[228] The work is beset by tensions between service to popular and practical needs and the interests of science, and the reviews indicated this. While aiming to follow the desires of his sponsor by attending to agricultural interests, Harris noted the limitations of the naturalist in addressing practical concerns. In this regard, near the end of the chapter on the Orthoptera, he confessed

After so much space has been devoted to an account of the ravages of grasshoppers and locusts, and to the descriptions of the insects themselves, perhaps it may be expected that the means of checking and destroying them should be fully explained. The naturalist, however, seldom has it in his power to put in practice the various remedies which his knowledge or experience may suggest. His proper province consists in examining the living objects about him with regard to their structure, their scientific arrangement, and their economy or history. In doing this, he opens to others the way to a successful course of experiments, the trial of which he is generally obliged to leave to those who are more favorably situated for their performance.[229]

In making this plea and explanation for the limitations of his work, Harris seems clearly to place himself among the naturalists.

A review of the work in the agricultural press, while appreciating its practical value, argued as well for its potential to elevate concern for natural science above "direct utility and immediate profit." This did not necessarily point to scientific ends, but to the hope that the vision of the farming population might be raised.[230] At least some agricultural editors recognized the need for farmers to key their interests toward science, and it was a reflection of Harris's reputation, derivative from the *Treatise,* when he was asked in 1847 to contribute to that end, as he was by the editor of a Philadelphia journal. The editor argued that "farmers, are very apt to be deficient—and I include myself among those who are most so—in matters of Natural History. The very objects with which we are perpetually surrounded are as perpetually neglected, without inquiry into their properties or habits."[231] Although he did not contribute to the journal, the sentiments for doing so are ones with which Harris would have agreed. Harris's first biographer, Thomas Wentworth Higginson, gave the case for the success of the *Treatise,* eloquently and simply, with the observation that "It is admitted by all who read this treatise that it is almost a model combination of the strictly scientific spirit with the clearest popular statement."[232]

The practical value of Harris's work, aside from the basic premise that control measures have to arise from knowledge of insect habits, has been the subject of debate and discussion among historians. In this regard, he was most often compared, among contemporaries, to the British entomologist John Curtis, whose *Farm Insects: Being the Natural History and Economy of the Insects Injurious to the Field Crops of Great Britain and Ireland . . . With Suggestions for Their Destruction* was published in Edinburgh and London in 1860. Curtis was noted as an entomologist and entomological illustrator, and his biographer quotes an 1846 letter where he seemingly apologizes for as well as explains his practical turn, "Finding that pure science led, like Poetry in the olden times, to starvation, I turned my attention to economic Science and engaged to write Reports for the Royal Agricultural Society on the Insects injurious to the Farmer." In fact his *Farm Insects* was based on his articles, published between 1841 and 1857 in the *Journal* of the society.[233] In a letter to Curtis in 1838, Harris mentioned that he had seen notice in the *Zoological Journal* that Curtis was "engaged in a work on British insects & plants" but he had not seen it and did not believe it was in any library in the area.[234] Exactly what work Harris was referring to is not clear, but in none of the letters seen between the two men was there any discussion of agricultural entomology. In 1846, however, Edward Doubleday wrote to Harris that he had lent Harris's work to Curtis and reported

that Curtis thought it was "the best book of the kind ever published."[235]

Curtis's *Farm Insects* had the advantage of including illustrations
from the start and by a man noted as an illustrator. However, his
biographer, George Ordish, claims that the book was the first "to be
written for the farmer," with the material arranged by farm crop
rather than by insect orders as in Harris's work.[236] In an earlier work,
Ordish noted these same sentiments in regard to Harris's work but
conceded that his "report did much good to patient readers for it
uncovered the life histories of many pests and prepared the way for
more practical approaches."[237] Asa Fitch, usually noted as Harris's
successor in applied entomology in the United States, made mention in 1851 of Curtis's "Observations on Insects affecting the Tur-
nip Crop" (from the *Journal of the Royal Agricultural Society*) as "a
perfect model for Essays of this kind,"[238] and in his own first report
as New York state entomologist in 1856 Fitch followed Curtis. He
wrote there, "It has been common in treatises upon economical entomology to arrange the several species in their scientific order. Although this mode of arrangement has its advantages, it presupposes
such an acquaintance with scientific entomology as but very few individuals in this country possess,"[239] undoubtedly thinking of Harris in
this context.

Entomologist and historian of entomology Leland O. Howard observed that Harris did not know very much about agriculture first
hand "and yet, with his careful accounts of the life histories of many
injurious insects, he laid a basis for much future work." Howard
noted that, while Harris was aware of the need for better remedies in
insect control, he was able to do only very limited experimentation
himself, in his own garden and relied largely on established procedures. These included shaking infested trees; hand picking the insects off the plant; enticing the insects with potatoes and poison
baits; pruning and burning twigs and shoots; use of substances such
as whitewash, tobacco water, soapsuds, and potash water; fumigation
with tobacco; forcing growth of plants through use of manure; and
soaking grain before planting.[240]

Although Harris was no agriculturist, he had much sympathy for
agricultural interests, and his own garden was said to have been one
of the delights of his life. His daughter reported that he always had
a garden in the several places where the family lived and rented land
to extend what was available to him. After moving into their new
home in 1843, Harris pursued his interests, as reported by the same
daughter many years later. "I don't know when my father bought a
lot of land on Cambridge St. to cultivate, nor why . . . This land

situated a little west of the present public library, was enclosed by a high board fence having gates and key and went by the name of 'the farm.' Here we delighted to go and spend some hours; sometimes I, the baby, would be helped on the way by a lift in the tipcart or wheelbarrow, trundled by one of Morris O'Conner's men employed to dig, weed and plant. Father would accompany the cavalcade and set the men to work, leaving us to go the Library."[241] His son, in discussing his father's study of cultivated squashes later in his life, referred to "his garden—next to his [insect] collection the delight of his life,"[242] and in that simple phrasing perhaps summarized the linkage, at a personal level, of Harris's interest in agriculture and the scientific study of entomology.

Scientific Work and Contributions: 1843–1856

As discussed earlier, Harris's life after 1842 was not what he expected it would be when he neared completion of his report on injurious insects for the state. Not only was his time more severely restricted, but the report itself gave him a visibility and reputation that put demands on him. His entomological work during the last phase of his life is more difficult to summarize because the large plans of his earlier years, even though only partially accomplished, are no longer present as signposts for his outlook and intentions. Though he discussed classification and the taxonomic relations of insects in his correspondence, his publication plans appear more ad hoc and more narrowly focused than in earlier years. They tended to relate to the habits of insects, their relations to agriculture, and to plans for a new edition of his *Treatise*. Nonetheless, in these endeavors he continued to view entomology in a broad and multifaceted way.

In the late 1820s, Nicholas M. Hentz referred to Harris as always following the system of the French entomologist Pierre-André Latreille (1762–1833),[243] who was known for his natural classification system, taking into account various elements of the insect body in deciding placement and arrangement.[244] Whether Harris was as closely tied to Latreille as Hentz states, by 1841 he was ready to follow "the best modern naturalists [who] are disposed to abandon Latreille's system" in favor of one that drew upon the characteristics of insects in all their life stages,[245] an approach that was fundamental to Harris's entomological thinking. His interest in exploring insect relations—as evidenced in his circular arrangements for the Bombyces (Lepidoptera), discussed above—won him the attention not only of his friend Edward Doubleday but also of Edward Newman (1801–76), also in England. Doubleday reported in May 1841 that he and Newman had spent "above an hour" discussing Harris's circles.[246] Newman, printer and naturalist, had issued a classification scheme based on the number seven that was circular in character, with a central group and with others surrounding it. Included in his 1835 *Grammar of Entomology*, the scheme had first been presented in a pamphlet in 1832 and was subjected to considerable criticism.[247]

In 1843, Newman published the second edition of his *The System of Nature: An Essay* (London: J. Van Voorst), which was dedicated to Harris.[248] In response, Harris took the opportunity to make some observations on Newman's views and to reveal something of his own philosophical reflections on the study of natural history:

> In a private course of lectures on Entomology, given to some of the students of the University, four years ago, I endeavored to explain your system, and made diagrams for the purpose, some of which still remain hanging in the room where our excellent friend, Mr. Doubleday, saw my collection of insects. I have often wished that you would combine in one work all you have published on the classification of insects & the characteristics of the groups. . . .
>
> If I cannot give an unqualified assent to all your views, I think them well worthy of attentive consideration & study. You have often very happily illustrated what before was obscure, & have pointed out some striking resemblances, or affinities as it is the fashion to call them. You have proved to my satisfaction the centrality of certain groups or types of form, combining some of the characteristics of the surrounding groups, together with a character peculiarly their own. This, it appears to me, must be the key to affinities, if such exist. That there are really 7 great & perfectly natural groups of insects, & that they approach each other, as you have represented, appears undeniable. Divide any one of them, and the parts lose their relative value when compared with the other groups. Whether there ever were, or ever will be, other equally natural groups of insects, & if so—how they can be connected with your circle is more than I can tell. It seems to me, however, upon taking a more extended view of nature, that living bodies are infinitely varied in structure, & were I to be required to say in one word, what is the system of nature, I should answer, variety. We see only a part of the series, the beginning and the end are lost to our view—we know only in part what is—we know but little of what has been, and we know nothing of what is to be. And yet, to form a perfect, philosophical system, or rather to trace out the whole plan of the Creator, we should have at once before us, all the living beings that ever have been, & ever will be created. Hence all our attempts to discover a natural system, either in Zoology or Botany, must fall far short of perfection.[249]

In his *Treatise*, Harris recognizes seven orders of insects that are "very generally adopted by naturalists," and the same were the basis of Newman's major insect arrangement.[250] It is likely that it is to this level of relations that he is referring when he assents to Newman's scheme.

It appears that Harris became increasingly skeptical about the ability to make claims for the reality of taxonomy, and in particular he came to question the nature of genera. In the early 1850s, he argued explicitly that too many genera were being constructed; part of his objection to dividing old into multiple new genera was the burden on the memory.[251] "I confess myself an unbeliever in a <u>natural system</u>, so called," he wrote, and elaborated by quoting his own words that reflect sentiments expressed in his letter to Newman quoted above:

> "We are not sufficiently aware, of what experience should have taught us, namely, that nature knows nothing of our systems, & that neither classes, nor orders, nor families, nor genera <u>were created</u>, but <u>species</u> only. When we reflect that the <u>living</u> races of animals & of plants constitute but a <u>part</u> of all that have been and all that may be,—shall we not be struck with the vanity of any attempt of mere finite intelligence to comprehend the extent or set bounds to the details of the <u>universal plan</u>. If I may presume to judge of the whole, from that small portion of it which comes within the scope of our limited faculties, I must confess that it presents itself to my mind as being a <u>system</u> of <u>infinite variation</u>, with a perfect adaptation of the species to the places they were designed to fill." So I wrote years ago,—and have had no reason to change my mind thereon.[252]

While he admitted somewhat later that his views "on the subject of genera, tribes, families, & the like, are at variance with most of those that prevail among naturalists, and are likely to find neither favor nor countenance with many of the latter," he had the consolation of "knowing . . . that they are shared by some who have attained a distinguished reputation in Europe, in the department of botany." But his attitude was not one of a proselytizer and he was content to coexist with a variety of views.[253]

During the last decade of his life, Harris interacted on a regular basis with two young entomologists who represented, in the next generation, the two directions in which insect studies were to go in the United States. It is a commentary on Harris's place in entomological history that he was on good and in certain ways mentoring terms with John Lawrence LeConte (1825–83), the recipient of the reflections quoted above, who was the son of John Eatton LeConte and the leading American scientific entomologist in the third quarter of the nineteenth century (specializing in the Coleoptera), and with Asa Fitch (1809–79), Harris's successor in applied entomology. It appears that Harris's correspondence with the two began in 1846.

Harris had apparently met young LeConte in 1842,[254] and the first letters known to have passed between them concerned Harris's transmittal of an insect and genealogical information on the Lawrence family.[255] Much of their correspondence was of a technical nature, especially the identification and relations of particular insects and insect groups, including specimens from different parts of the country, in the course of which Harris referred to his collection, his experience, and his familiarity with the printed literature. LeConte, in presenting his views on the nature, identification, and relations of genera and other groups, including the difficulties posed by the lack of fossil evidence for insects, said "I am glad you wrote to me on the subject for I have no one here [in New York] to discuss these matters with, although I am extremely anxious to compare my views with those of other naturalists."[256] LeConte also credited Harris with a special knowledge based on his studies of life histories of insects, commenting late in their relationship on the difficulty he was having with one Coleoptera genus and "wish very much that you who have observed the habits so carefully would give us an account of the limits of variation of our more abundant species."[257]

J. L. LeConte was a prolific author and during the years of his correspondence with Harris he produced a number of entomological papers,[258] while Harris produced relatively little that was not, by title, agriculturally oriented.[259] Although unable to undertake any sustained work on his own, it can be argued that Harris contributed to the advancement of scientific entomology during the last decade of his life, at least in part, through his assistance to the younger entomologist. LeConte sympathized with Harris's predicament, writing in 1847: "If your Amara are to be published soon I could put off the publication of my Amaroides, & Harpalidae . . . I should be sorry to interfere with anything you have done, & as my time is of very little value to me, I might as well occupy myself with some other family in the mean time; while yours is so much taken up with your duties as Librarian, that it would take you some time to prepare another paper for publication. Let me know your views on this matter speedily."[260] As late as 1851, LeConte did not have a copy of Harris's *Treatise,* for at that time he asked for one to be sent to him,[261] which suggests that LeConte was reliant on Harris because of his general reputation as entomologist rather than because of his best known publication.

Harris also developed a continuing relationship with Asa Fitch, and it seems to have been richer for both men than his relationship with LeConte, because their interests were closer to the central concerns of Harris's later years. Fitch visited Harris in August 1845, and this apparently was their first contact. Although Fitch spent the night at Harris's house and had access to his library, his visit coincided with the Harvard commencement and Harris was unable to devote much attention to him.[262] Their early correspondence related to the Hessian and the wheat fly (species of *Cecidomyia*) on which Fitch was working when they first met.[263] Harris disagreed with some aspects of Fitch's account of the insects.[264] On the other hand, they easily agreed on the definition of a species, defending the "old, definite criterion of reproduction." In Fitch's estimation, "though we may be regarded as antiquated in our notions, and behind the times," he fully expected eventual vindication.[265] In response Harris wrote that "I agree with you relative to the value and real and permanent condition of species. As to genera, and other groups of species—your views may perhaps change when you get to be as old as I am. The longer I live, the less confidence I have in such groups being natural, or that systems are any other than contrivances of man to assist his limited faculties."[266] They later found themselves in agreement, as well, regarding "the excessive multiplication of the genera and species."[267]

In August 1852, Harris noted the special relationship that he and Fitch had and the way in which he depended on their letters:

> I had begun to get <u>impatient</u> at not hearing from you. There is nobody here with whom I can confer on such subjects as engage our attention; none for whose opinion, in the present case, I care a <u>straw</u>; none, by discussing with whom any doubtful question in Entomology, I can feel that my own wits are sharpened, my ideas brightened, and my observations quickened and rendered more acute. What if the other party does not happen to agree with me upon any disputed point, I feel that discussion with such does me good, and may promote the cause of science; I may convince, or be convinced, and to this end keep my mind open to conviction.[268]

Through the later years of Harris's life, one of the subjects of special interest was the Hessian fly and related dipterous insects destructive to wheat and to some other grains. Of the former, Sorensen writes that it "has the distinction of being the first insect pest to plague American farm crops on a massive scale" and became particularly destructive in the midwestern part of the country during the years 1844 to 1847.[269] Although Harris was in fact dependent on Edward

C. Herrick and others for much of his Hessian fly information in the *Treatise,* that work took on an authority such that he was consulted on the subject.

Among those with whom he became involved in this way was Philadelphian Margaretta Hare Morris, who had been studying insects injurious to wheat for some years and had reached contested views on the Hessian fly. Morris opened a correspondence with Harris in 1843[270] and when she later became involved in a controversy with Fitch over the fly, she called on Harris for his opinion on the matter.[271] In the revision of his *Treatise,* Harris tentatively concurred with the conclusion that Morris had by then reached, that her insect was not the Hessian fly but a near relative.[272] While counseling Morris later about an error she had made in confusing the larvae of a Lepidoptera insect for a Coleoptera, Harris revealed much about his perceptions of his own strengths as an entomologist, his relations to Fitch and LeConte, and his mentoring and corrective role. On that occasion he wrote,

> Now do not be disconcerted on account of these mistakes. There are very few persons, even among naturalists, who have a competent knowledge of the larvae & pupae of insects. Such knowledge forms a study by itself, and is to be obtained only by long experience. Some naturalists, of good attainments in other departments, have made as great mistakes about larvae. Dr. Fitch, for example, was as much mistaken about the joint worm, which he thought to be the larva of a *Cecidomyia.* As soon as I came to examine it carefully with a magnifier, I saw at once that it was a Hymenopterous & not a Dipterous larvae. So. Dr. Leconte mistook the larva of Corydalis cornutus for that of a beetle, . . . He also made a report on another larva to the Scientific Association, which had it been published, would have cast disgrace on American science. Luckily, I arrested the report before being published, & pointed out to Dr. Leconte his error. Do not think I mean to arrogate too much to myself. The study of the larvae & transformations of insects dates with me from boyhood, and I am now 57 years old. It would, indeed, be strange if I could not speak with some confidence on the subject.[273]

In the early 1850s, Harris was able to direct the outcome of his various studies of the Hessian fly and its allies toward revision of his *Treatise.* He wrote to British entomologist John O. Westwood about a copy of the second edition of the *Treatise* he had sent and pointed out sections of the revision that he considered to be particularly notable, including the transformations of the *Cecidomyians* (the genus to which the Hessian fly belonged). In this regard, he was particularly critical of Fitch.

In my anxiety to do full justice to writers who have treated upon the history of these insects, I have given to Dr. Fitch rather more credit than belongs to him,—certainly more than he deserves at my hand, considering the somewhat flippant stile [*sic*] of some of his remarks upon statements made in my former account of these insects. . . . Until Dr. Fitch's attention was particularly directed by my inquiries, to the particular mode in which the change from the larva to the pupa was affected, he did not realize that there was anything therein more than common, & certainly had no clear conception of the nature of the process, till the receipt of my letter touching this point. I might very well have claimed the credit of making the discovery myself; not directly, but indirectly through him; and in my Treatise it is for the first time distinctly stated & described.[274]

This rather unbecoming display of immodesty, whether true or not, must have been intended to curry personal favor from a leading foreign entomologist but also reveals a rivalry Harris may have felt with the rising American agricultural entomologist.

Although Harris seems to have done little organized work in entomology in the later phase of his career, certain projects were begun but, at one point he complained, "It is very discouraging that every attempt of mine to return to entomological studies should be met by constant disappointments, or by such hindrances as interfere with any thing like a systematic attention to the subject," adding that "Desultory & spasmodic efforts afford little satisfaction & profit either to myself or to others."[275] One project that Harris undertook in this period and which was carried to completion was his assistance to Louis Agassiz (now at Harvard) in working up the results of the latter's exploration of the northern shore of Lake Superior in the summer of 1848.[276] Agassiz turned over to him all of the insects collected on the Lake Superior survey and offered him any duplicates that he desired.[277] Harris prepared the description of the Lepidoptera for Agassiz's published report on the expedition,[278] an event that Augustus Radcliffe Grote later took as evidence of the esteem that Agassiz felt for Harris as an entomologist.[279]

Around the same time as the Lake Superior work, a potentially much larger project presented itself, but there is no evidence that he ever became involved. In September 1848, Charles Wilkes, commander of the United States Exploring Expedition to the Pacific in 1838 to 1842, contacted Harris to ask if he would undertake description of the Expedition insects for publication (excluding the Coleoptera, which were already done). Wilkes was authorized to pay, but he was unable to estimate the amount of work the collections would require, in effect inviting Harris to inspect the insects in Washington. Wilkes did indicate that the insect collections were not as extensive as he would have hoped. As further enticement, he pointed out that other naturalists known to Harris already were engaged in work for the Expedition, including Augustus A. Gould (mollusca and shells), Asa Gray and Edward Tuckerman (botany), and Louis Agassiz (ichthyology). Shortly after Wilkes's letter was written, Harris got one from Charles Pickering, who had been the chief zoologist with the Wilkes Expedition, urging Harris to undertake the work.[280] Harris's response to the invitation has not been seen. There is no evidence, however, that he ever worked on the project and presumably that was by his own choice. A volume on insects from the expedition, in fact, never appeared. Historian of the expedition William Stanton noted that John L. LeConte's intended volume (on the Coleoptera) was written but never published, and, overall, it appears that the insects were described by various individuals but the results were not published in an official expedition volume.[281] A project that would have substantially increased Harris's entomological range, therefore, must stand only as a token of the place that he occupied in the entomological community of his time.

In 1850, the Massachusetts legislature called for a reprinting of Harris's *Treatise* on injurious insects. As noted earlier, Harris had thoughts of a second edition of his *Treatise* as soon as the first was completed in 1842. By 1846, the first edition supply was nearly gone and he was making a concerted effort toward a new edition, this time with illustrations.[282] He had canvassed some one hundred correspondents "requesting specimens of destructive insects, & notes on their habits & repeated appeals have been made through agricultural papers, to persons interested in cultivation, for communications on the subject." These efforts did not provoke much response, however, but, by "my own good fortune," he had been able to gather some new materials.[283] His work on the new edition was encouraged by his Mississippi agriculturist friend, Thomas Affleck, who thought that "it will take well, & do a vast amount of good," but counseled Harris to take pains with the work and the illustrations, for "it is only the Farmer, of some education, who will buy it, & he will prefer it so."[284] Affleck, an agricultural writer himself, asserted that "Such a work as you can prepare, if got up in proper manner, will sell well. It is wanted. The time has come for it."[285] Asa Fitch also expressed delight at Harris's plans for the revision, hoping to see, "among other additions, . . . our other nocturnal Lepidoptera worked out with an approximation to that completeness with which our Bombyces have already been presented."[286] The next year, Fitch explained how knowledge of Harris's plans had persuaded him to revise his own, noting that he had been led to abandon his intention to contribute on injurious insects for the agricultural volume that Ebenezer Emmons was preparing as part of the work of the New York Natural History Survey. Giving up his plans to write on the injurious insects of New York, as unnecessary duplication of effort, in favor of Harris's revision of his *Treatise,* Fitch announced his intention to devote his time to work on the Homoptera.[287] Fitch later counseled that Harris ought to append his catalogue of Massachusetts insects to the work (presumably referring to the catalogue that had appeared in Edward Hitchcock's state report in 1835), noting that "although the science has progressed considerably since it was published, it is still not the best, merely, but the only Catalogue of our Insects which we possess, and therefore continues to this day a document of much value, and one which will be of service to almost everyone who purchases your book."[288]

In passing its resolves in 1850 for a new edition of the *Treatise,* the Massachusetts legislature provided for two thousand copies and authorized Harris to take copyright in the work. For the effort, Harris was to be paid one hundred and fifty dollars and receive two hundred of the copies for his own distribution.[289] The secretary of state, who was charged with seeing the project carried out, urged Harris to make haste, noting how great the demand was for it.[290] Pleading the demands of his position in the library, Harris was given dispensation to proceed at his own convenient pace,[291] but the legislature was impatient.[292] After nearly two years from the time of the legislative resolve authorizing the new edition, Harris "continued to give to it such time, as my other duties & my health would permit," promising completion during the coming winter.[293]

In spite of Harris's earlier announced intention, the legislature made no provision for including illustrations in the revision. In fact Harris did not think the local talent was available to carry out such work, and what they could provide was likely to be so costly as to place inclusion of figures "beyond the limits of any reasonable appropriation."[294] Though apparently not a prospect for the revised edition of his *Treatise,* Harris did correspond about that time with Townend Glover. Glover was a person of leisure with artistic talents and interest in entomology who was then looking for a place in American science and would become a notable illustrator of insects.[295] Glover wrote for information on an insect (found on his pear trees) that he did not find in Harris's *Treatise* and included a sketch of the insect; he also mentioned sending Harris "a few of the first fruits of that conversation [in Boston] in the shape of engravings on stone, wh [*sic*] you suggested might be done." Glover solicited Harris's opinion of his engravings and stated that "if you approve, I will endeavour to take some lessons."[296] Harris responded apparently with approval and with the proposal that Glover make himself available to prepare illustrations for him. Harris, however, could not have meant this for the revision of his *Treatise*, which was then either done or would be soon.[297] Glover indicated interest in Harris's proposition but explained that he was then too busy to make any decisions. He hoped to visit Harris on his return from Washington, if in fact he went.[298] There is no evidence that Glover ever did any illustrative work for Harris, however, and in fact his visit to Washington led to employment there which occupied much of the remainder of his life.

Harris does not often comment directly on the relative value of work in systematics, applied entomology, and popular science. Thus, an apologetic letter composed in 1854, following transmittal of a copy of the second edition of his *Treatise* to English entomologist, John Obadiah Westwood, is particularly revealing. It has to be read with some reservations, however, in that Westwood was one of the most distinguished British entomologists of his generation,[299] and Harris, as he composed the letter, may have had certain expectations of the reception it would receive in that quarter. His ambivalence in addressing the issue is seen also in the fact that, in the passage in question, all except the first sentence was bracketed on the available draft of the letter, and in fact may never have been sent to Westwood. Harris wrote:

> I am aware that the greater part of this work will be of little or no interest to you. The very homely stile [*sic*] in which it is written may be thought beneath the dignity of a naturalist. In excuse for this, it may be stated that the book was designed mainly for the use of farmers & gardeners, and other persons engaged in rural pursuits, who stand in need of a plain account of our most common destructive insects, arranged in such a way as to convey to the readers some notion also of the classification & natural history of these insects. That the book has to some extent, supplied a want that had been long felt here may be inferred from the fact that two editions of it have already been disposed of.[300]

Westwood's reply, if one was received, has not been discovered.

In 1855, Harris was faced with a proposition from C. M. Saxton, a New York agricultural publisher, who wrote to explain that he had requests for Harris's *Treatise* which he could not meet and suggested that he reprint the work for Harris. Saxton thought it might be "very useful to the great mass of farmers"—though admitting the sales probably would be limited.[301] Three years after the second edition of the *Treatise* had been published, Harris still had seventy copies unsold.[302] Apparently he decided on the frugal course to protect the value of the unsold copies by not agreeing to a reprinting of his work. At any rate, by the time Saxton received a letter from Harris,[303] accompanied by a copy of the work, the publisher had been to Washington and had spoken with Townend Glover, who was working on illustrations of insects in the South. Saxton had proposed that he expand his work to cover the entire country and suggested collaboration with Harris (who would do the textual part), which would produce a work of considerable value to the public. As for Harris's own work, Saxton noted that it was limited to New England, and he was satisfied that its reprinting by the state would meet the apparent need. He did agree to take ten copies for advertisement and sale on Harris's behalf.[304]

SCIENTIFIC WORK AND CONTRIBUTIONS: BOTANY

Harris had an interest in botany early in his life, and his son said the subject "possessed for him always a peculiar charm."[305] His ongoing concern with the life stages of insects, and thus their relations to their food supply, naturally kept his botanical interests active, though secondary. L. O. Howard thought Harris was an accomplished botanist, a fact that made his entomological work "broader and sounder."[306] In spite of these interests, however, Harris published little on the subject, which included the "List of Native Plants Discovered Growing Near Boston, the Present Season, in a letter read before the Massachusetts Horticultural Society" (*Magazine of Horticulture* 6 no.7 [July 1840]: 245–47).

In Harris's important letter to John Lowell in 1827 on the status and needs in natural history in the United States, he observed that botanists should study species common to the older and the American continents "to ascertain the means by which they were introduced here, and trace out the causes of the difference observable in their sensible qualities."[307] During the last years of his life, Harris undertook botanical studies that seem to follow this early advice. His studies of squashes and pumpkins (order Cucurbitaceae) in fact combined literary or historical studies with the natural history of the plants. In 1850, Harris wrote to Charles Sumner for help.

There are some interesting questions, connected with American History & Botany, not satisfactorily settled. One of them has arisen from what is stated of the plants, cultivated by the New England Indians in the account of the voyage of De Mons along our coast in 1605, where only the common French names of the plants are given. To approximate to a determination of the species, we must know what the French plants are, that resemble them so nearly as to justify navigators in applying their names to the American species. If we had a few seeds of these French plants, we could, by cultivation, have specimens for comparison, and thus perhaps be able to settle the question.

Harris had heard that Sumner's brother, with whom he was not acquainted, was in Paris and asked if he could procure some seeds for him to aid in his study.[308]

A year later, he wrote to a friend and explained that due to a recent illness, he had not been able to get out and therefore had only books for resources. Taking that opportunity for study, he began a more deliberate investigation of "the history & nomenclature of the pumpkins and squashes noticed by early European visitors to North & South America," borrowing from the library what he described as "numerous ponderous tomes of Greek & Latin." His letter was addressed to the office of the secretary of the state at Albany, where he had previously sent a paper on pumpkins and squash. Now he asked that "Mr. Johnson" not print it, explaining that he was writing up the results of his work, "with citation of authorities," for submission to the *American Journal of Science* and he "should be sorry to have it anticipated."[309]

Harris's efforts to gather information on the squashes and pumpkins included correspondence with his old Harvard colleague, Thomas Nuttall, now living in England. He hoped to elicit recollections of the plants that were cultivated by native Americans encountered by Nuttall during his earlier travels in America, but Nuttall recalled little that would be helpful.[310] In 1853, Harris wrote to Pennsylvania botanist William Darlington to inquire about procuring some seeds from the pumpkin *Cucurbita maxima,* and noted, "I have been cultivating & studying all the principal species & varieties of this genus that were accessible to me. I have satisfied myself, on historical grounds, that they are all originally American."[311]

Harris's son wrote of him that "For several seasons his garden . . . was filled with squashes, pumpkins, and gourds of every conceivable shape, size, and color. Seed came to him from all quarters of the globe, and their products formed a collection as unique as it was interesting."[312] As with other projects, this one does not seem to have come to full fruition. The article under preparation for *American Journal of Science* never appeared. In 1851, he published "On the History and Nomenclature of Some Cultivated Vegetables (abstract)" (*Proceedings of the American Association for the Advancement of Science* 5 [1851]: 180–82), which related to the project. His bibliography includes reference to short articles on custard squash, acorn squash, and pumpkins and squashes published in the *New England Farmer* in 1851 and 1852.[313] He did have more ambitious plans, however, for at his death he left a substantive though incomplete illustrated manuscript on the order Cucurbitaceae.[314]

CONTRIBUTIONS AND INTERESTS OUTSIDE SCIENCE

While Harris's primary focus was on science, he also had other interests and concerns that he followed less methodically or with less devotion. Describing his tastes and abilities as librarian, his son noted, "He possessed, in addition to his extensive knowledge of many branches of the natural sciences, a keen love for and appreciation of the fine arts, was an interested student in geography and history, a good classical scholar, and a fair mathematician."[315] The early history of the American (especially New England) settlement was a particular interest. His studies in this area were not carried out in the continuing and methodical manner characteristic of his work in entomology, but they did make it possible, as one of his obituaries noted, for him to give informed assistance to others who followed a more dedicated course.[316] His son characterized his antiquarian interests as more in the vein of a student than writer.[317] Some of his historical research related to Harvard University and may have been part of his official duties as librarian; for example, in 1847 he undertook an investigation of Harvard Yard (including trips to the city clerk's office) and reported his progress to the university treasurer.[318] He also assisted President Josiah Quincy with the appendix of historical documents to his *History of Harvard University* (1840)—in 1846, somewhat belatedly it seems, Quincy sent him a copy of the two volume work and tried to get Harris to change his mind and accept payment for the work he had done.[319] Harris seems to have published very little of his historical and antiquarian research, the first apparently being an article on the early Cambridge printer Stephen Daye in the Boston *Courier* (July 29, 1847).[320] In 1846, he came into possession of some of the papers of Harvard's first president, Henry Dunster, and through them he was able to determine some of the facts relating to the Reverend J. Glover,[321] who had a part in the introduction of printing into America. When the editor of the *Boston Saturday Rambler* wrote to Harris and asked him to write on agricultural topics, Harris responded that he already was engaged in a similar arrangement to write in exchange for subscriptions with several other agricultural papers and would, therefore, be unable to accept the arrangement. However, he did offer an article he had written on Glover, which the *Rambler* published in two parts, on September 1 and 8, 1849.[322] He also published on the life of his medical mentor and father-in-law, "The Late Dr. [Amos] Holbrook" (*Boston Medical and Surgical Journal* 26, no.23 [July 13, 1842]: 358–60).

He had an active interest in genealogy and published "Notes on the Josselyn Family, of Massachusetts" in the *New England Historical and Genealogical Register* (2, no.3 [July, 1848]: 306–10). He was "deputed" to write a history of the Mason family (the family of his paternal grandmother) and in 1849 reported it nearly done.[323] Like so many of Harris's projects, however, the history was never published, although, on his death, he left a substantial manuscript on the Hugh Mason family of Watertown.[324] His son Edward, who came to share his father's entomological interests, noted the common qualities that Thaddeus William Harris brought to these studies: "He seemed to possess an instinct for unerringly tracing a genealogy, and the wandering individual who was encountered in his search was almost as surely and correctly assigned his proper place in the line of family descent, as was the wary insect allotted in his cabinet its true order and genus."[325] Harris became an honorary member of the Historical Society of Pennsylvania in 1846 and a member of the Massachusetts Historical Society in 1848, both likely indications of his developing interests in historical studies in his later years.[326]

Harris's talent for drawing and interest in architecture were manifested in various ways.[327] He prepared preliminary plans for his own home in the early 1840s, as discussed elsewhere.[328] His abilities in drawing were an integral part of his entomological studies—for example, his drawings of the veins of the wings of insects for purposes of identification and classification, and of larvae as a supplement to his notes and preservation of the specimen.[329] He also undertook more formal preparations for the projected work on Lepidoptera which was to be published by the American Academy of Arts and Sciences.[330] His very first published paper, on the salt-marsh caterpillar in 1823, included a plate with nine figures.[331] However, Harris seems to have realized his limitations and the superior skills of others such as Nicholas M. Hentz, John Abbot, and Townend Glover.[332]

LIBRARIANSHIP

During the majority of his adult life—the twenty-five years from 1831 until his death—Harris's primary concern and activity was not entomology or historical study but the administration of an academic library. The historical appraisal of this phase of his life has been somewhat ambivalent, due, it seems, to an awareness among his contemporaries that his chief interest was in natural history, and among those who followed by the more commanding impression left by his immediate successor, John Langdon Sibley. Sibley, who had been an assistant in the library from the early 1840s and worked closely with Harris, gave his own assessment: "I was a witness to the economy, the exactness, and the fidelity which characterized his administration."[333] Harris's nineteenth-century biographer agreed, noting that "he kept his official records with exquisite accuracy, and described his methods to other librarians as lovingly as if he were describing a chrysalis."[334] One of the first things that Harris did to regularize administration was to prepare an annual report to the Corporation (at the suggestion of the president), beginning at the end of his first year.[335]

During the first ten years of his tenure, it appears that Harris worked largely alone, with some help from a janitor.[336] The staffing problem, in fact, never was settled for Harris. His report for 1853 lamented that "not only are the Librarian & Assistant Librarian [Sibley] taxed to utmost in the performance of their increasing duties, but it has become apparent to them that, without additional help & additional funds, it will be impossible to carry on the work satisfactorily . . . It is confidently believed that there is no library in the country which, in proportion to its size, is carried on with such limited means, or which has a greater claim to be provided with ample funds."[337] Harris's library duties increased greatly especially after 1841 and the move to the new Gore Hall. In 1844 he wrote to his English friend Edward Doubleday, "Your account of your own labors, cares & anxieties in the [British] museum, seem to me an echo of my own, in the public Library of the University, which now for two years has added a double burden to my before overtaxed power of mind & body."[338] Assistant librarian Sibley reiterated this situation after Harris's death, noting that "a greatly increased responsibility [made] . . . constantly greater calls on his industry and diligence."[339] Harris reviewed the development of his responsibilities and gave a detailed account of his duties as they had evolved to 1845, in writing to one of his neglected entomological correspondents:

The administration of a library of nearly 50,000 volumes is no sinecure. My duties in this library have been constantly & greatly increasing for several years, by the increase of persons using the books, of those who visit the library from curiosity, & by large additions recently made, & still making, to the collection. In the year 1842 a subscription of 20,000 dollars was raised to buy books. A list of books to be bought was immediately prepared, orders were sent abroad, and the books began to come in during the years 1843 & 1844. In the last year alone, there were received, by purchase or gift, 3,000 volumes, & 2,500 pamphlets, all of which have been catalogued since their receipt, analysed & classified according to subjects, the analysis & class written down in books for the purpose, the books have been mostly arranged on the shelves, marked to their places & entered on the shelf catalogues. In the term time from 800 to 1000 volumes are borrowed <u>weekly</u> from the library, & as many returned, & in vacations, on every Monday, from 3 to 4 hundred. A regular entry of titles & places of books borrowed has to be kept, & the return of them to be noted therein also, with a receipt from the borrower. These duties, with all others belonging to the place, are discharged by myself and one assistant. We are employed in term-time every day, excepting, perhaps, part of Saturday, & Sunday, 10 or 11 hours in summer, & 7 & 8 hours in winter; & we have had to work also during a great part of the vacations. You can judge how much leisure & ability will be left to me for any other employment.[340]

By 1840 Harris had underway a plan, "long ago required by the College Laws," for an undergraduate collection that would be placed in the reading room of the new library, thus making a selection of books more accessible to students.[341] Harris wrote to the president to promote this scheme and, knowing the time required to make the selection, wanted the faculty to act on it soon, although it is not clear whether Harris or the faculty were to make the choices. Harris thought that 1,000 to 1,500 volumes would be appropriate for the student's library. It appears that most of the books were to be selected from titles already in the library, although Harris did recommend the acquisition of several sets of books and noted that some other modern works would be required. Harris gave the president ten categories of "books most used by students," namely "1 English Works of Fiction, Novels, &c.; 2. Dramatic Works in English 3. Reviews & Magazines; 4. Ancient & Modern History; 5. Voyages & Travels; 6. Biographical Works; 7. Literary Essays; 8. English Poetry 9. Works of Fiction, Drama, Poetry &c. in French, German, Italian, & Spanish; 10. Natural History."[342] In July 1841, the corporation ordered the plan to be carried out and some 2,000 volumes were selected or purchased.[343]

As early as 1835, Harris had begun plans for moving to a new library building.[344] The actual move into Gore Hall took place in the summer of 1841.[345] Sibley, who had been assisting with occasional tasks, joined the staff at that time.[346] The move itself took eleven days, beginning on July 19, the books being transported in long boxes "in the order in which they stood on the shelves at the time of the last examination, and were placed in the new library, on shelves designated beforehand, nearly in the same order as before." However, about 8,000 volumes previously shelved as miscellaneous were assigned to subject classes in the new building, and adjustments had to be made in the corresponding catalogues.[347] The physical arrangement of the books was an ongoing concern, even in the new building, and by the late 1840s Harris recognized the need to "classify and arrange the books anew," because of the pressures of books added since moving into the new library building. In fact this task had begun several years before.[348] By the time the next annual report appeared, the rearrangement of the alcoves had been completed and their new catalogues had been prepared.[349]

The library grew substantially during Harris's tenure, adding about 36,000 volumes to the 33,000 that were there when he began in 1831, plus an equal number of pamphlets.[350] This growth took place chiefly in conjunction with the move to Gore Hall and in the years thereafter. During the 1830s until 1843, the average number of volumes added in a year was less than one thousand, whereas in the next five years (1844–48) the average was about 2,400 volumes (3,645 in 1844 alone).[351] During the later period of his librarianship, from 1849 to 1855, the number of volumes added ranged from 724 to 2,887 with an average of about 1,600.[352] Harris notes the number of received items that were gifts, which were substantial.[353] Assistant librarian John Langdon Sibley, who succeeded Harris as head of the Harvard library, is especially known for his aggressive collection of publications of all sorts, and these efforts began when he worked with Harris, who acknowledged Sibley's success in his 1847 annual report.[354] When he began his manuscript journal soon after taking charge of the library in 1856, Sibley especially mentioned that during his fifteen years as Harris's assistant, he had added "donations to the library from 15000 to 20000 pamphlets and not less than 7000 volumes which but for his [Sibley's] personal exertions would never have been added."[355]

Harris worked hard at collection development and on the prospect of finding adequate funds to keep the library balanced across fields of study and up to date. Sibley, in his review of Harris's librarianship, noted that during his years more than $35,000 was expended on the purchase of books. But this never became regularized. The most systematic effort took place at the time of the Gore Hall move when (as related in the letter quoted above) some

$21,000 was raised for the purchase of books.[356] In 1838, in contemplation of the new library building and the opportunities it offered, Harris reviewed collection needs in his annual report.

> This appears to be a proper time to bring before you, Gentlemen, the expediency of providing means for supplying the deficiencies in the library. . . . No pains should be spared to increase the department of American History, deservedly the most important & most valuable portion of our library; and the departments of Intellectual Philosophy, Modern Theology, and Natural History, which at present are but poorly supplied, ought to be brought, at least, to an equity with the other departments. No addition seems to be necessary at this time to the branches of Philology, of Classical, & of General Literature, nor to that of the History of Continental Europe, in all of which the library is richly supplied. A few works on Mathematics & Natural Philosophy, & but a few, may be necessary. Some more biographical & bibliographical works will be exceedingly useful, &, indeed, seem to be indispensable; for your librarian was obliged to borrow them of individuals when employed in making the supplementary catalogue.

Harris's comments on the requirements in natural history were quoted and summarized in the previous chapter. He continued with what he hoped would be the occasion and the incentive to carry out a more systematic buying program for the library along the lines he had shown. He also pointed out to the Corporation some of the practical reasons not to delay.

> I understand that great additions to the library are contemplated when the books shall be removed to Gore Hall; but as some time must elapse before orders can be executed & the books received, even if the orders should be transmitted to Europe immediately, I conceive that it may be thought expedient to ascertain the deficiencies of the library forth with, & that it is not now too early to begin to provide the means for supplying them. I know not in what way you contemplate to obtain these means; but, if they are to be procured by subscription & by personal solicitation, it cannot, surely, be too soon to make known the actual condition of the library . . .

He ends this appeal with a personal statement combining deference to the university's higher authorities and a claim for special knowledge: "Gentlemen, if the opinion of your librarian, who, of all persons, ought best (from the position in which you have placed him) to know the wants of those who frequent the Public Library of the University, & if the advice of one who offers it also as a naturalist . . . be of any value, you will not think that he has urged upon you, with unbecoming zeal, the importance of devising & obtaining, without delay, the means to increase the Public Library, so that it may recover and retain the rank which it has heretofore held in public estimation."[357] As indicated above, funds were procured, which led to an acceleration in the growth of the library and an ongoing increase in the duties of the librarian. By 1846, the library held 50,955 volumes, not counting unbound pamphlets and books then with the binder.[358]

The large subscription of 1842 did not bring sustained support. In 1850, funds available to the library, all intended for specific purposes, yielded an income of $450.[359] Two years later, Harris attempted to make a case for government aid to the library, apparently having in mind general support in addition to funds for book purchases.

> It is a subject of much regret & solicitude that the college is not provided with any permanent fund for defraying the expenses of the Library, which continue to be a charge upon the general income of the institution . . . A permanent fund of one hundred thousand dollars would not be more than enough to place & keep the Library on a proper foundation. In no better way, than by endowing the Library with such a fund, could the legislature of Massachusetts carry out that provision of the Constitution, which makes it "the duty of legislatures and magistrates in all future periods of this Commonwealth, to cherish the interests of literature & the sciences, and all seminaries of them, especially the University of Cambridge." In addition to this consideration, the Library has a peculiar claim to the bounty of the Commonwealth; for the destruction of the first Library was caused by the occupation of the Hall by the General Court.[360]

Harris's reach was ambitious but unrealistic. Interestingly, his suggestion of state support came on the heels of a constitutional crisis for the university the previous year that at first threatened the institution with greater state control and then resulted in the reduction of representation by the Commonwealth on the Board of Overseers.[361]

During Harris's tenure, the library catalogue underwent a change that reflected the increasing size of the collection but also manifested a degree of ingenuity on Harris's part that linked methodologically with his entomological studies. Harris's predecessor, Benjamin Peirce, had printed *A Catalogue of the Library of Harvard University in Cambridge, Massachusetts* (1830–31) in three volumes, arranged alphabetically by author with a subject index,[362] and Harris's early efforts were devoted to production of the *First Supplement*

(Cambridge: C. Folsom, 1834) of 260 pages, which covered additions to September 1, 1833. As the title indicates, further supplements were planned, but they never appeared.[363]

Harris's nineteenth-century biographer and former student, T. W. Higginson, relates how it was only after Harris died, and when he had access to his papers, that he understood how Harris achieved such a command of entomological classification. Higginson's description of Harris's manuscript indexes and compilations, from published entomological works, was quoted in the previous chapter. Harris used similar methods in preparing catalogues of his insects, which Higginson described as based "on the plan of the card-catalogue now used in libraries, upon uniform pieces of paper, three or four inches square, which he afterwards tied in bundles, and carefully labelled. Each card contained the name of the insect with synonymes [sic] and authorities, and the number it bore in his catalogue,—but no description." Higginson concluded that much of this work was carried out in 1837.[364] In 1852, Harris himself described something of his methods, noting that, in preparing to transcribe his insect catalogues, which "have become so filled up with notes & interlineations," he planned first to compile "an alphabetical reference catalogue to all Say's descriptions, to facilitate the finding of his names & descriptions. Having adopted the plan of writing each name on a separate piece of paper, about 3 by 2 inches, it will be easy to arrange them hereafter either alphabetically or scientifically."[365]

Harris's work in the library and in entomology required item identification and relationship, and he appears to have moved between his bibliographical and natural history work with similar methods. He has been credited with some degree of innovation on the bibliographical side, but much of the procedure for compiling information on slips or other writing pieces was in use by others, and perhaps fairly widely. For example, Say's widow in 1835 mentioned that her husband wrote his insect descriptions on small pieces of paper because he "found them more convenient for the printer."[366] Furthermore, one of Harris's predecessors as librarian, William Croswell, in 1812, began preparations for a new library catalogue by cutting up the previous printed catalogue entries into strips, which were arranged in the preferred order and pasted in volumes along with additions to the library likewise entered on slips. It is in the context of Harris's own use of various procedures for the organization of information, and similar practices by others at the time, that he has been credited with the first recorded proposal for a card catalogue in an American library.[367]

Harris's credit for the idea of the card catalogue is based on a proposal in his annual report of 1840, where he recommended "That the Corporation should authorize a slip catalogue to be made, consisting of the title of every work in the library on pieces of card $6^1/_2$ inches long & $1^1/_2$ inch wide; such catalogue being much wanted when books are arranged for the annual examinations to indicate missing books, & would also be extremely useful in facilitating the rearrangement of books in the new library, & would serve for various other useful purposes hereafter."[368] Unfortunately, if there was a response to the idea from the Corporation, it is not known. A letter from Harris to Treasurer Samuel A. Eliot in 1848, however, sheds light on how the idea apparently stood in limbo for some years before it was acted on. Harris wrote,

> The plan for cataloguing mentioned by you, I proposed to Mr. Ward [Eliot's predecessor as treasurer] more than ten years ago, suggesting the use of thin smooth cards for the titles of each work. He gave me no encouragement or authority for doing the work in this way; & since that time, when I have spoken of it to Mr. Sibley [the Assistant Librarian], the latter has made all sorts of objections to it, & therefore I have gone on as I had begun. Mr. Folsom,[369] however, upon my suggestion, has adopted the plan, & likes it much. As the catalogue has now been made upon paper written only on one side, so as to be cut into strips, it is a subject for consideration when these strips shall have been assorted, compared & corrected, whether they shall be pasted on sheets in alphabetical order, or put upon cards singly. If the catalogue is not to be printed, it will have to be in constant use in the manuscript form;—& it then becomes a question whether, for general use, single cards or sheets of titles will be best. The latter may be made up into volumes; the former must be kept in files.[370]

The reference to "slip catalogue" in Harris's original proposal in 1840 suggests the procedural and historical relations to Croswell's earlier cut-and-paste methods, where Harris's plan would use cards rather than sheets as the physical unit. The letter to Eliot reveals, however, that the idea of a sheet or card catalogue still was under discussion in 1848. At any rate, the card catalogue of Harris's design was in existence as of 1848, which seems to indicate that Eliot concurred with the idea and it went forth at that time.[371] The meaning of Harris's statement in his letter, however, where he wrote "The plan for cataloguing mentioned by you," raises an interesting question as to whether, in fact, Eliot suggested the card catalogue as an idea original to him and not with knowledge of Harris's 1840 recommendation. The answer to this question has not been discovered.

Sometime before 1850, the bibliographical entries were written directly on the cards rather than pasting entries from some other source, which appears to indicate that it had become a primary means of control over the collection rather than derivative or supplementary. Whatever the specific historical details and chronology, Harris's catalogue (using cards measuring $9^1/2 \times 2^1/2$ inches) was an instrument intended for the use of the library staff and not by readers.[372] The day of the printed catalogue was passing, and the card catalogue was the instrument for the future. After 1834, there would be no more printed catalogues for the Harvard Library, a fact that reflected general trends for American libraries in the middle and later years of the nineteenth century.[373] As a local event, the introduction of the card catalogue as a viable substitute resulted not simply as a logical outcome of a general historical movement, however, but from the innovative ideas of Harris, in which he was able to draw not only on his activities as librarian but his related and passionate interest in entomology.

Harris's father, Rev. Thaddeus Mason Harris, built the Dorchester, Massachusetts, family home where Harris grew up. Known as Mt. Ida, it was sold in 1840 and razed in 1916. Courtesy of the Dorchester Historical Society.

Thomas Say (1787–1834), the first important student of American insects, gave early aid and advice to Harris. Courtesy of the Ernst Mayr Library of the Museum of Comparative Zoology, Harvard University.

Nicholas Marcellus Hentz (1797–1856) was Harris's earliest colleague in entomology. His best known work was on American spiders. Courtesy of the Ernst Mayr Library of the Museum of Comparative Zoology, Harvard University.

Charles Pickering

Charles Pickering (1805–1878), botanist and anthropologist, was one of Harris's important naturalist friends. Pickering helped to facilitate Harris's access to the active scientific community in Philadelphia. Courtesy of the Harvard University Archives.

Edward Doubleday (1810–1849), assistant in the British Museum in charge of moths and butterflies, was a close professional friend of Harris and instrumental in encouraging and facilitating his study of the Lepidoptera in the late 1830s and early 1840s. Medallion in plaster by Bernard Smith, 1844. Courtesy of the Natural History Museum, London.

Gore Hall opened in 1841, during Harris's tenure as Harvard librarian. It was the first separate structure at the university dedicated to housing the library collections. Courtesy of the Harvard University Archives.

Asa Fitch (1809–1879) was appointed New York state entomologist in 1854 and was the leading investigator of agricultural pests following Harris. The two men were acquainted from the mid-1840s. Their ongoing correspondence allowed Harris to participate on a personal level in preparing the next generation of economic entomologists. Courtesy of the Ernst Mayr Library of the Museum of Comparative Zoology, Harvard University.

John Lawrence LeConte (1825–1883) began studying insects in his teens and became the leading Coleopterist in the generation following Harris. They began corresponding when LeConte was still a very young man and their relationship kept Harris involved in scientific entomology in a time when he was unable to do any sustained entomological investigation. Courtesy of the Ernst Mayr Library of the Museum of Comparative Zoology, Harvard University.

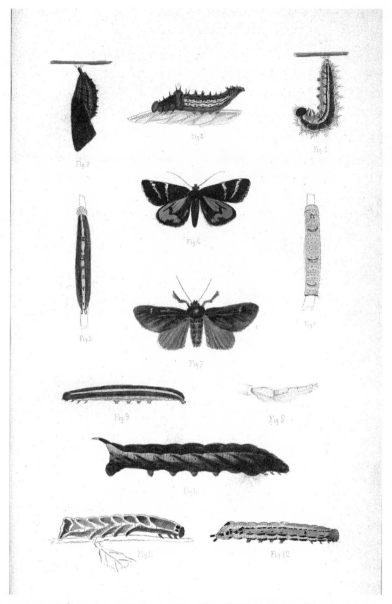

From: *Entomological Correspondence of Thaddeus William Harris, M.D.*, edited by Samuel H. Scudder (1869). Plate 1 [original plate is in color]. Figures are based on drawings by Harris, with the exception of no. 7 (moth in the middle of the plate) and no. 8 (tarsi or foot, at right below no. 7). Courtesy of the Ernst Mayr Library of the Museum of Comparative Zoology, Harvard University.

Harris's gravesite in the Cambridge Cemetery, also showing the headstone of his wife Catherine Holbrook Harris. Five of their children are buried in the same lot. Photograph courtesy of Priscilla J. Elliott.

4

In the Community of Science

Ethics and Expectations: A General View

JAMES ELLSWORTH DEKAY, WITH WHOM HARRIS WAS IN CONTACT REGARD-
ing the insects for the Natural History Survey of New York, had a
cynical and suspicious perspective on the character of the scientist,
writing to Harris, "I have waited for some time to send the package
by some individual not addicted to Natural History, I hear such sad
accounts of Naturalists in general & of Entomologists in particular
that I am afraid to trust them."[1] In Harris's contrary view, however,
dedication and selfless interest was an adhesive as well as a lubricant
that made the workings of science possible in spite of obstacles. For
example, when he was trying to recruit another New York state resident to collect insects for him (and was able to offer only the naming of insects in return), he hoped that love of science would be a
sufficient motivation, a condition "which I have generally found to
be joined to a liberal & generous disposition."[2]

Later in the century, Augustus Radcliffe Grote used Harris as an
ideal against which to reflect the bad conduct of naturalists in his
own day, contending that "the fault of science is its quarrelsome-
ness, its egotism, its belief, not in humanity, but in itself." By contrast, and using Harris's writings (especially the *Treatise*) to make his
point, Grote argued that

> He never misrepresents anyone, nor does he abuse the confidence
> which the State of Massachusetts has reposed in him by circulating asper-
> sions upon the work of others . . . He is extremely careful of other
> people's reputation, neither ridiculing ignorance nor concealing inde-
> pendent discovery. . . . He makes room for the unknown, for that at least
> which he does not know. . . . Dr. Harris is throughout unambitious of
> himself, intent only on bringing out entomological facts in pursuance of
> his duty. His report, therefore, nowhere reads like autobiography; it
> treats solely of the doings of insects, . . .[3]

His nineteenth-century biographer remembered that Harris once mentioned the British scientist Robert Brown as an ideal of the scientific reputation, in Higginson's words, "supreme among botanists, unknown even by name to all the world beside."[4] Harris was not without concern for personal status, however, noting early in his career, when he was working on the Faunula Insectorum Bostoniensis that the "book will either establish my reputation or promulgate my disgrace."[5] But his ambitions were also for American science, for the development and promotion of entomology, the fulfillment of popular interests and the meeting of practical (i.e., agricultural) needs, and the establishment of rules for the practice of natural history and the just assignment of recognition for work done.

ROLE AND STRATEGIES IN SCIENCE AND IN RELATION TO THE PUBLIC

Harris's scientific work was never simply the pursuit of knowledge for its own sake. It *was* that, but his genuine interest in popularization as a strategy for promoting his subject, and his life-long interest in the applications of entomological knowledge for practical purposes, preclude easy characterization of his life as a scientist. He struggled with these issues in the 1820s as he tried to establish himself as a scientist while practicing medicine, and looked about for ways to generate interest and therefore social support for entomological studies. His Faunula Insectorum Bostoniensis (discussed in chapter 3) was intended to be useful and to meet especially the needs of young students.[6] Addressing popular interests was not a strategy that Harris reserved for himself. Thomas Say was the country's most notable contemporary scientific entomologist and, taking his cue from Say himself, Harris urged him to end his descriptive *American Entomology,* which Say was issuing in parts with colored illustrations, and suggested a less expensive work that would be more "widely circulated, and would make the science of Entomology popular & respectable in this country."[7]

Harris's motivations for publishing in the agricultural press are considered below, but the reasons he gave for contributing to the *New England Farmer* were similar to those given for work on the Faunula, to help young entomologists in the region and to assist in the development of American science.[8] In criticizing an 1835 article by B. Hale Ives (1806–37) published in the *American Gardener's Magazine,*[9] Harris argued that such journals are sent to Europe "where they may fall into the hands of men of science." Concern for scientific reputation, therefore, required that work be carefully done and that authors "adopt a correct nomenclature."[10] Ives wrote in response that the "lack of a Manual for the young student in Entomology is a great drawback, and deeply as I love to wander in the fields & study this beautiful portion of God's creation, I yet am obliged to forego the additional pleasure of being able to give each its specific name,"[11] sentiments that must have reinforced Harris's feelings about the value of elementary works as a way to serve both the young investigator and the reputation of American science as well.

Publication aside, Harris employed an altruistic argument in support of his collecting and arranging insects, which he saw as a means as well as an acceptable end for his scientific efforts. In 1837 he wrote to Charles J. Ward, an Ohio collector (whom he was prepared to instruct and assist in exchange for specimens):

> My object in making a collection, and for this purpose, asking the aid of my friends, has not been merely personal gratification: it has been my desire to add something to the cause of science in this country, where entomology has found few admirers, and still fewer teachers. . . . [E]ven should death surprise me before the result of my labors are before the public, I shall leave an extensive, well arranged, & named collection.[12]

A modern historian has characterized the early economic entomologists as motivated by a "flagrant altruism," by a drive to be useful (a characteristic shared with other educated citizens, in this case, to save agriculture), and in the process to serve their own interests by rising above a generally negative image that stigmatized the student of insects.[13] Harris was at the vanguard of the movement, but this portrait of the economic entomologist describes him only in a very limited sense. The connection to agriculture was part of his entomological program from the outset, a genuine interest that drew on and incorporated his great interest in the life histories of insects. He had a love of his subject and that was the underlying motivation, while he desired sufficient time to prosecute his studies. He was genuinely interested in the popularization of entomology, which he considered a "never failing source of the most rational enjoyment."[14] He also had the hope and expectation that, from more widespread knowledge and appreciation of the subject would come greater support (implying more help and support for his own investigations),[15] and, of course, the outcome would also benefit farmers and horticulturists.

PUBLISHING STRATEGIES

Harris's conception of his place in both the general and the scientific communities, and how they interrelated during his lifetime, are revealed as much through his publishing strategies and rationale as in any other way. Harris's situation is particularly instructive among his contemporaries as it shows the relations between the agricultural press and other publishing outlets that served a scientific community in the process of differentiation and professionalization. A large percentage of his published output appeared in agricultural journals, and yet, as discussed in the previous chapter, his orientation and aspirations came to transcend that limitation but never went entirely beyond it. Harris understood the difference between the two modes of publishing and sensed the tensions that it posed for his self-conception as a naturalist.

In 1828, Harris wrote to this naturalist friend Charles Pickering about his project to contribute to the *New England Farmer*, at the request of its editor. His plan was to publish there, "once a fortnight," material on insects, including their habits and descriptions of new species.[16] But he intended to send the descriptions alone to the Academy of Natural Sciences of Philadelphia beforehand with the hope that members of the Academy (including Pickering) may be able to tell him whether some already had been published elsewhere and therefore were not new to science. His problem was that the Academy's *Journal* had suspended publication and yet he felt the need to get his descriptions of new insects into print as soon as possible. Describing the *New England Farmer* as "a paper having limited circulation only, & not ranking among the scientific Journals of the day," he had every expectation that the descriptions, being first read to the Academy, could be republished in its *Journal* when it was resumed. Informing Pickering of his intention to send to him, on an ongoing basis, descriptions of supposed new insects to be read to the Academy in advance of their publication in the *Farmer*, he asked to be informed if the publication committee objected to his plan or if the *Journal* was not to be continued. In this case, he would likely send his descriptions to the *American Journal of Science*.[17] In reply, Pickering reported that he had not communicated Harris's paper to the Academy because no one there except himself was interested in entomology and, furthermore, its *Journal* was not to reappear soon. He suggested that Harris send his descriptions to the *American Journal of Science* instead, since they would be little noticed in the *New England Farmer*. For the larger question of publication policy, Pickering advised that, when (or if) the Academy resumed publication of its *Journal,* there probably would be some objection to including Harris's insects if they had been published previously.[18] Harris later informed Pickering that, on account of the precariousness of the Academy's *Journal,* he had decided not to submit new insects to it.[19]

Shortly after the letter to Pickering, Harris also wrote to Nicholas Marcellus Hentz about his publication plans, which contained something of his motivations for his entomological work in general. There he admitted that he was

> aware that the "New England Farmer" is not likely to be much circulated among men of science, and therefore will not be considered the best authority; but it is a convenient vehicle at present; and, such is the ambition of European entomologists to anticipate Americans, that I willingly yield to the solicitations of several friends in publishing what may possibly contain many new species; and, in doing so, I am not actuated so much by personal considerations as by a desire to aid several young entomologists in this vicinity, and by the wish to promote American science in general: *pro patria.* The "Farmer" is taken at New Harmony, and will therefore come under the eye of Prof. Say: it is my intention, after these descriptions shall have undergone his rigid scrutiny, to republish them, either by themselves, or in some respectable scientific journal.[20]

Thomas Say, although having relocated from Philadelphia to New Harmony, Indiana, was expected to play the role of judge for Harris's entomological work that the members of the Academy at Philadelphia were not able to perform.

A possible explanation why he did not follow through on his idea of publishing in Silliman's *Journal*[21] was his preparation of the catalogue of his collection that appeared in Edward Hitchcock's report on the geology and natural history of Massachusetts in 1833. In 1831, he was appointed to the Harvard librarianship and, from the slightness of his preserved correspondence during the period, he appears to have engaged in little entomological activity in that year.[22] Early in the following year, however, he was looking for an outlet for his scientific work. His interest in finding a scientific journal that could publish his descriptions of insects was communicated to the Boston Society of Natural History, which undertook to investigate the possibility of establishing a journal and reported back favorably on the undertaking.[23] The Society's *Boston Journal of Natural History* began issuance in 1834,[24] although Harris never contributed very extensively to it. In the same year that he interested the Boston society in establishing a scientific journal as an outlet for his descriptive work, he also delivered the fourth anniversary address to the Massachusetts Horticultural Society on the relations of insects and plants. It was considered a highlight in the society's early history and was described many years late as having included "more information of practical value to cultivators, than any other address ever delivered before the Society, being a summary such as had never previously been published of all that was then known in regard to insects injurious to vegetation here, and the best means of preventing their ravages, by the most accurate and thorough student of the subject which the country then possessed."[25] Published separately by the society in 1832, the address also was serialized in the *New England Farmer* the following year.[26] These two activities of 1832 indicate that entomological writings of different genre belonged in publishing outlets appropriate to their intent, and in particular that scientific work belonged in a scientific journal.[27]

For the remainder of the 1830s, Harris did not publish directly in the *New England Farmer,* his most frequent outlet in the 1820s. Instead he published chiefly in the serial publications of the Boston, Hartford, and Essex County natural history societies, the *American Journal of Science,* and in Massachusetts state reports. After his report on injurious insects in 1841, however, the *Farmer* (and other agricultural journals) again became a frequent place for his writings. In broad outline, therefore, Harris's places of publication tended to map the status of his entomological career. In the 1820s, when he had few outlets, he attempted to combine agricultural and scientific work in the *New England Farmer;* during the 1830s, when he was more concentrated on scientific work, he sought venues other than the agricultural press; finally, after the publication of his Massachusetts report on injurious insects and the pressure of other work that prevented sustained research in systematics, he reverted to the agricultural press.[28]

Naming Insects: Standards for Practice and the Politics of Natural History

Harris developed strong views on practices among naturalists, and these concerns were an important component of his orientation and actions within the national and international scientific community. It was in the practice of naming insects that Harris was particularly outspoken and, while much of it was directed at Europeans, he also was willing to critique American naturalists, referring in 1826 to Say's preference for several names, over those having what Harris considered as greater authorities, as "vicious nomenclature."[29] One of Harris's primary points was that only names accompanied by a published description should be given recognition, and the name given by the first naturalist to produce such an accessible description should have priority.[30] Historian Gordon R. McOuat, writing on British practice in the nineteenth century, has discussed the social role of names among naturalists. The conservative elements argued the Linnaean view that species names need not have meaning in themselves, which was a function of the description. The radicals among British naturalists, tending to be provincials, urged the view that names should have meaning in themselves and not simply arbitrary labels.[31] Though Harris tended to the conservative view on priority and form of names,[32] in certain respects his views on the signification of insect names resembled those of the radicals, although no systematic statement on this has been discovered. One important point about which Harris felt strongly, and which was referred to in the previous chapter, is that he followed the rule of Linnaeus that, whenever possible, the specific name of an insect should be the genera on which the insect feeds,[33] thus showing meaning through relationship in nature.

In the question of communal practices and acceptance of priority of names Harris was a nationalist in science, motivated largely by a sense of hurt that Americans had suffered at the hands of Europeans and by a desire to promote scientific study in the United States. Harris was not alone among Americans during the Jeffersonian period in his feeling that European scientists were depriving Americans of the privilege of describing and naming their own natural products.[34] He did follow a consistent approach to the problem, however, and spoke out on ways to counteract European incursions. He identified the problem as derivative particularly from the practices of French naturalists, reflecting, in part, his own New England Federalist roots. Although he was dependent on Europeans not only for their published descriptions of American insects but also for their personal assistance in identifying specimens in his cabinet, he developed the strategy of assigning names to insects in his collection which he could not identify, before sending them abroad. In 1829 he outlined his practice to his friend Nicholas Marcellus Hentz, explaining that he had given names to some "doubtful or new coleoptera" before sending them away. "If they were previously known to Entomologists, Mr. Faldermann [at St. Petersburg] will inform me, & if new my name will stand, & Europeans be anticipated. By taking this precaution we may safely send away undetermined species, & be benefitted by the remarks of our correspondents, without committing ourselves in print."[35] Having worked out a strategy for himself, he was critical of others, such as John Eatton LeConte, who seemed to send insects to Europe freely without taking adequate precautions to protect American interests.[36] Outlining the terms by which he could accept assistance and exchange of insects with LeConte, Harris stated that for nondescripts he would require that LeConte retain the names he had given them and gave his own pledge to publish them as soon as possible "in some of our public Journals, that you may have the sanction of these Journals for adopting the names."[37] LeConte agreed to the terms.[38]

In 1830, Harris wrote at length to his naturalist friend Charles Pickering, who did not think there was much chance of preventing the export of American insects to Europe, since there already were many more there than in American collections.[39] Signifying agreement with a statement by Pickering, Harris then outlined the essentials of his own approach to the best practice and protocol in the naming of natural history objects. He urged that

> (even if others act differently) let us not in future send away any insect without a specific name: if we can find a legitimate one it should be adopted, if not we ought to affix a name ourselves & establish it, as soon as possible, by publishing a description of the species. On reflection you will be convinced that a correct description is the only legitimate authority for any name, & the first describer is alone responsible for it; &, further, that bare catalogues, without descriptions, cannot designate nor be any authority for specific names. How are we to know whether an insect has or has not a specific name, if it has not been described? Without doubt it is the description alone which designates the name, and publishes it to the world.—Look at Dejean's "Species"[40]—you will find that many American insects have more than one name. Almost every Entomologist in Europe gives such a name as he pleases to a nondescript, and how are we to decide between them?—Say answers, you will answer, all of us must answer—the first describer. Of what consequence is it to us that there are hundreds (or thousands if you please) of undescribed American insects in the cabinets of Europe, all of them with names, & many with half a dozen different ones? Does this give us any better knowledge of what these insects are?

Harris asks further, "Is science to remain stationary in this country, till the good people of Europe shall think proper to give us descriptions of our natural productions?" Following Pickering's lead he reveals, "Ever since receiving your advice on the subject I have forborne sending abroad any insects which I had not determined except half a dozen, perhaps, to which I gave my own names." He then connects his own ambitions and interests to the program for priority recognition and the protection and promotion of American entomology: "It is not my wish to interfere with <u>American</u> Entomologists. I always intended to confine myself to the description of insects found in New England, & particularly those in the vicinity of Boston. If any of my countrymen will undertake, <u>immediately</u>, any part of the task it shall be relinquished in their favour. By all means let us anticipate foreigners."[41]

In his expectation of help from Thomas Say in 1834 in naming insects for the revision of his catalogue for Edward Hitchcock's Massachusetts report, Harris appealed to the need to reflect positively on American science.[42] After Say's early death, Harris labored to promote the fate of his fellow American's work among European naturalists, and in particular to establish and preserve the priority of names assigned by Say, "<u>names sanctioned by the descriptions which accompany them.</u>"[43] As always, he was particularly critical of the French. To the English entomologist John O. Westwood, Harris made the point that Say was little regarded in France, where works in English and German generally are ignored. Observing that some two-thirds of Say's insects were described in the *Journal of the Academy of Natural Sciences of Philadelphia* and in the *Transactions of the American Philosophical Society,* he suggested, "If you will take the trouble of comparing the dates & descriptions there found with the same in Cte Dejean's Species gen des Coleopt., where very many of our insects are redescribed under other names, you will not be surprised that Mr. Say's friends should feel indignant."[44] Westwood made some complaint about the scarcity and cost of American entomological works in England, a situation that Harris countered with his belief that American publications were regularly sent to individuals or societies in Great Britain and regretted the possibility that they might not reach their destination.[45]

The competition with Europeans took on a new immediacy for Harris in late 1838 when he met the traveling English entomologist Edward Doubleday and realized the extent to which the Englishman had collected American Lepidoptera. Although a close and supportive friendship developed with Doubleday, Harris expressed some ambivalence in regard to his own and the interests of American natural history, as suggested in the previous chapter. Reporting to his friend Edward C. Herrick shortly after Doubleday's visit, he noted the many insects that Doubleday had collected in the United States. "A great many were new & I suppose that he & Newman will describe them[.] I am sorry for this, but cannot help it; & hope that entomologists will rise up among ourselves to prevent foreign collectors & describers from carrying off the honors which we ought to earn & claim for ourselves."[46] But Harris's suggestion to Doubleday that he and Edward Newman "come here, [and] become citizens of Massachusetts," where he would make available his undescribed insects,[47] suggests Harris's willingness to go to great lengths to promote the study of native insects when done by naturalists resident in America.

Historian W. Conner Sorensen writes that, while American entomologists in the 1830s tended not to follow Harris's lead, by the next decade the important Entomological Society of Pennsylvania had adopted his practice of not sending unnamed insects to Europe. Sorensen calls it a "protective embargo."[48] When the society wrote to Harris in 1842 to enroll his cooperation in the preparation of a catalogue of American Coleoptera, they explained that it would omit "all mere catalogue names; and we intend to pursue the same course that our ornithologists & conchologists followed, in naming and characterizing all the species in our collections which we cannot identify from the works in our possession; convinced that as long as this course is unadopted, so long will entomology be without a name among us."[49] This was similar to the strategy that Harris adopted in the 1820s when he published in the *New England Farmer*, although subsequently he had expressed some possible reservations about the practice.[50] The Pennsylvania entomologists also had their reservations, noting that "We will doubtless inflict many synonyms upon the science, & meet with much opposition from abroad; but we must either pursue this independent course, or quit the study."[51] To the enterprise Harris counseled patience. Frederick E. Melsheimer, president of the Entomological Society, had been charged with compilation of the catalogue and Harris wrote: "If you will allow me to advise, in such a matter, I would recommend that the publication of your Catalogue should not be hurried; for many opportunities to add to it, & to clear up the synonymy will be likely to occur; & if you should allow it to lay over some months, or even a year, you will not regret the delay."[52] Sorensen discusses the question of strategy as a dilemma for the society between the establishment of priority of names on the one hand, and the quality of the product and possible critique by European naturalists on the other. John Lawrence LeConte, an important presence in the society, followed Harris's approach and promoted a deliberate rather than hasty completion of the catalogue.[53] As with Harris, however, LeConte's hesitancy to go into publication did not indicate any lack of concern for the interests of American naturalists. In 1845, the young LeConte published a statement later referred to as "America's Entomological Declaration of Independence,"[54] and his language is strikingly similar to Harris's in his correspondence from a decade and a half earlier.

In the summer of 1849, Harris and Agassiz engaged in a brief epistolary discussion about names in natural history (regrettably, only Agassiz's letters have been found). Harris provided Agassiz with criticism of his *Nomenclator Zoologicus* (issued in twenty-six parts; Soloduri, 1842–46), a compilation and arrangement of names of genera in zoology, with reference to the place of first publication and the family of which the genus is part.[55] Agassiz hoped sometime to do a revision of the work and assured Harris that he would then employ his advice and assistance regarding the insects, explaining that "When I prepared the list of genera of Lepidoptera I had only had your book for a very short time I had not yet studied it, but simply extracted the new names from it. In the future it shall be my guide in these matters, with your personal advice, if you will have the kindness to continue it to me [*sic*]."[56] The next day Agassiz sent his "rather bulky" notes on Lepidoptera. They were arranged alphabetically, but he gave Harris permission to rearrange the leafs according to system and asked Harris to offer his detailed opinion regarding the classification of Lepidoptera, meaning "the value of larger divisions, the secondary groups being well characterized in your Report."[57] Harris wrote to Agassiz shortly thereafter on the matter of nomenclature, to which Agassiz graciously replied, "I always derive equally pleasure and instruction from my intercourse with you. I have read with great interest your remarks upon nomenclature. Some points are presented by you in a light in which they are scarcely ever looked at and I acknowledge with the truest spirit of liberality and justice." But he could not agree with all that Harris wrote and hoped they might be able to discuss their differences.[58]

In addition to procedural matters on nomenclature, Harris apparently also divulged to Agassiz his long-held critical attitude toward European naturalists. The Swiss-born naturalist and new arrival to the American scientific community took the opportunity to try to reconcile the two, but he could not have been aware of Harris's particular francophobic perspective. "It may be that there has been an unpleasant rivalry between the english and american and that the latter have suffered in that quarter; but never did any such thing occur in continental Europe," mentioning several individuals as examples of the good relationships and high regard in which American naturalists were held, including Say who "is estimed [*sic*] as an eminent Zoologist wherever Nat. Hist. is cultivated." Agassiz then said, "let me ask you whether you have read Erichson's reviews of your Treatise on the Insects injurious to vegetation? If you have you must then know that your own countrymen never did you the justice which you so deservedly received at the hands of my german friend.

But you need not mind the petty jealousy of those who have no higher aim than naming species; you have done enough to feel that you stand above such a level."[59]

Harris was largely a man of words, which he sent out to perform their actions as best they could. It is not apparent that he took a leading role in the social organization for reform of nomenclature and related matters in natural history, but, in addition to practicing what he preached, in his own way he attempted to promote the organization that would bring about needed change. In 1838, Harris wrote to the British entomologist John O. Westwood about the problem of synonyms in natural history and noted that a "voice from this western wilderness" would have little effect on the problem, which must be addressed at the centers of science in Europe. "I can only hope that my humble efforts in the cause of truth & justice" in time will be supported by European naturalists. "I beg leave therefore to suggest to you, my dear Sir, who are imbued with the true spirit of this [cause?], that this subject might be submitted & discussed, with the best effect, at the meetings of the British Association for the Advancement of Science [BAAS], & at the reunion of the German Naturalists. By these distinguished bodies, or by a still more general convocation of naturalists, the whole nomenclature of Zoology & Botany could be fairly discussed & finally settled." Harris suggests beginning with a rendering of justice to Linnaeus, thereafter revising and purging the names of his contemporaries and immediate successors. As for the names derived from naturalists after the early post-Linnaean period, Harris suggests that separate commissions might be appointed for the various divisions of natural history, and then at a "subsequent convocation the whole subject could be submitted to the assembled naturalists of the world, & brought to a final adjudication."[60]

How or if Westwood responded to this suggestion is not determined. In the mid-1820s, British entomologist William Kirby had suggested the appointment of a committee within the Zoological Club to deal with nomenclature questions, and this was done but came to nothing.[61] When a reform movement in natural history developed in Great Britain in the early 1840s it was not the systematic review of names, as Harris seemed to suggest, but a writing of rules (although ideas similar to Harris's had circulated in Britain in the 1830s). The lead for reform was taken by Hugh Edwin Strickland (1811–53), within the BAAS beginning in 1841. A committee on rules of nomenclature was appointed by the BAAS in 1842; Westwood was a member. Although not formally adopted by the BAAS,

he rules as drawn up were printed in the *Report* of its annual meet-
ng and took on authority by association.[62]

Within the American natural history community, Harris experi-
nced a sense of isolation and therefore ineffectiveness. In 1839, he
vrote that "If the entomologists in this country would only communicate freely & liberally with each other, & would unite their efforts
o promote the advance of the science, something might be accom-
»lished; but, at present, few as we are in number, far removed from
·ach other, & each one forwarding the science without reference to
»ne great & common object, we have only the mortification to find
hat our efforts are disregarded both abroad & at home."[63] Like
»ther naturalists, he was looking for larger organized involvement
vith the American scientific community. Harris wrote the same year
o John G. Morris, "I have long thought that a congress of natural-
sts, convened annually in some [of?] our principal cities, would
;reatly tend to promote science in this country, & have proposed it
o some of my friends. But most of us are too poor to meet the ex-
)enses of such an undertaking without the cooperation & direct aid
»f rich amateurs." He mentioned that John Collins Warren had sug-
;ested the idea recently to the American Philosophical Society, but
hat body rejected it and nothing came of the proposition.[64] Perhaps
t was poverty, or that he was distracted by his increasing library du-
ies, but when organization took place leading to the formation of
he Association of American Geologists and Naturalists in 1840, Har-
is does not appear to have taken any part. The organization, how-
:ver, began among geologists and the name initially was confined
o that science. The name change that broadened the scope of the
ssociation to include naturalists came at its meeting in Boston in
ate April 1842, and yet no mention of the fact has been found in
Harris's correspondence.[65] The month of April 1842, of course, was
a period of great personal trauma for Harris, having suffered the
oss of the Fisher Professorship to Asa Gray, and his father's death
»n April 3. Whether Harris was a member of the association and
vhether he took any part when it met again in Boston in 1847 has
not been determined.[66]

At the 1845 meeting of the Association of American Geologists
and Naturalists in New Haven, a slightly revised version of the British
Association rules on nomenclature was adopted and recommended
o American workers. Harris is unlikely to have found objection to
hem, and his chief complaint in his correspondence over the years
vas addressed in rule 3a, which is presented as one of the limits on
he rule of priority: "Names given to species or groups unaccompa-
nied by published characteristic descriptions, should yield place to

the earliest name accompanying such descriptions." The committee
appended a commentary on the rule, where they stated that "It ha
been customary with some naturalists to give names to species ir
their cabinets, or in a published catalogue, and on this ground, tc
claim authority for such names. This should not be allowed." These
are words that could have been written by Harris some years earlier
The committee also gave guidance on the type of publication tha
would qualify under the rule: "Neither is it sufficient that the de
scription appear in a public newspaper, or in a journal not generally
known for its scientific character, or in language so brief and in
definite that the object cannot be recognized by it," also words tha
generally reflect Harris's own sentiments and practice. Among the
committee that drew up the rules were several friends and col
leagues of Harris in the Boston scientific community—Amos Binney
Augustus A. Gould, and David H. Storer, as well as his New Haver
friend and sometime entomologist Edward C. Herrick.[67] It is a curi
ous fact that Harris appears to have had no part in the deliberations
regarding these rules in spite of his early interest in such contro
measures. Whether he would have played a greater part a decade
earlier, before his scientific activity began to wane in the face of
heavy obligations to the library, is hard to say. It does suggest, how
ever, that Harris was more effective in formulating principles than
he was in acting cooperatively to bring about change.

NETWORK AND COMMUNITY

The study of entomology in the early nineteenth century was com
plex, as natural history in general was, driven by communal and per
sonal motives, the drive to advance knowledge, to justify or to find
social value for the enterprise. But the exchange of information—in
the form of written (printed) texts, specimens (both new and identi
fied), and expert opinion and advice—was the means by which the
process worked. Harris's search for resources for his study has been
considered in an earlier chapter. But none of that activity was for
solitary work. Rather, it was part of an international economy, proba
bly especially robust for entomology because of the great number of
species and the small size of the specimens that were easily ex
changed. Harris entered fully into the business. He expressed the
situation in 1838 in a letter to New York entomologist Edwin Will
cox, whose assistance he desired, noting that duties in the university,
a large family, and a small income made it impossible to make "ex
cursions into the country & neighboring states. [Therefore] I am

nable to do much towards increasing my collection by my own ex-
rtions. For the future, I must depend upon the generosity of my
·iends for the addition of species to my cabinet, and for supplying
1e with the means of continuing exchanges with my foreign corre-
pondents." Having explained his own needs, he felt his past ser-
ices justified calling in a collective debt due to him. "Having always
een willing & ready, on my part, to give my services to all who
1ought proper to apply to me for aid in the determination of our
1sects, it is not asking too much of my American friends to help me
1 return, especially if they will consider how tedious & trying it is to
1e eyes & patience carefully to examine & compare with my own
1sects, one by one, all the insects sent to me for their names." By
1is time in his career, Harris was in a position to offer his services
1 naming insects sent to him, which gave him a medium of ex-
hange in the absence of physical specimens. This was a position
arned by hard work over a period of time and did not come all at
nce, for in the same letter he told Willcox, with some pride, that,
ven though Thomas Say (who had died in 1834) provided him with
 number of names for his specimens, I "had it in my power, occa-
ionally, to assist even my lamented friend."[68]

The international exchange of natural history specimens was
·remised on mutual obligation. For example, expatriate American
·otanist Francis Boott wrote from London to Harris that if he
·ished to establish an exchange with John Curtis, he should send
·urtis a selection of his most rare specimens and Curtis would then
·eel compelled to send others in return.[69] Harris himself indicated
hat the same etiquette governed domestic relations, writing to John
·atton LeConte in 1838 that

> Mr. W. [Willcox] informs me . . . that you have some insects for me,
> which I flatter myself are intended to complete the unfinished exchange
> begun by you more than eight years ago, and which I have been hoping
> to receive from you ever since that time. Forgive me for reminding you
> of the promises, contained in your letters of 12th & 31st March 1830,
> and for stating that for the 231 insects which I had then sent to you I
> have received only 45 individuals from you in return.

·he fact that Harris felt it possible to assert his interests in this way,
·riting politely, characteristically specific, but also uncharacteristi-
ally aggressive, to an established entomologist of reputation, indi-
ates that his reminder of LeConte's obligation was within
·cceptable behavior among naturalists. This is underscored by the
act that, in the same letter, Harris asks, even "at the risk of being

charged with presumption," to borrow LeConte's collection of Cole optera,[70] a request he would not have endangered by stepping out side the bounds of expected mutuality in regard to the exchange o specimens.

The sale of insects was a part of the economy of entomology, bu it seems, at least for Harris, to have been not only a minor part bu one not entirely acceptable or respectable. Part of the reason ma have been Harris's own economic status. He wrote to the Britisl entomologist George Samouelle in 1837, acknowledging that Sam ouelle had insects for sale but expressing the hope that an exchange could be arranged instead.[71] In the event, Harris made contact witl John Curtis, to whom he expressed pleasure at the prospect of ar exchange of insects, suggesting that this was a preferred relationshij to buying insects from Samouelle. In response to the expected re ceipt of insects from Curtis, Harris wrote that "You may depend or receiving an equivalent in insects of this country."[72] Harris wrote to Nicholas Marcellus Hentz in 1838 to solicit his aid in the develop ment of the Boston Society of Natural History's collection of Lepi doptera and to generally assist Harris with his own work on tha order (a focus of his efforts at the time). "I am authorized to offer you the sixteenth of a dollar [apiece?] for the butterflies, 5 cents for the sphinges & large moths, & 4 cents each for the small moths,' but noted that if the rate was not suitable, Hentz should name hi own. Harris added that he would purchase larvae and pupae on hi own, since the Society had no apparent interest in these forms.[7] Hentz replied to both the personal and institutional requests, but ir different ways, writing, "I am so willing to aid you that since yester day I have prepared two nets to collect Lepidoptera; and I will no only send you what I catch, but am very willing to assist you with my pencil, if you wish it. This I mean to do for nothing save friendship' sake. As to the gathering of Lepidoptera for the Society, I accep their offer." In the spirit of the enterprise, however, Hentz did sug gest that, in exchange for insects sent, he would welcome foreigr insects of the orders Coleoptera and Hymenoptera.[74] Though Harri might have been willing, if necessary, to pay for insect specimens, i was not something he engaged in himself, writing to Edwin Willcox that he had "never sold insects . . . but have given away thousands without any return whatever."[75]

The traffic in insects and entomological information was a com plex network, involving hierarchies of shifting relationships. The individuals with whom Harris interacted ranged from the acknowl edged experts to the more than fifteen American collectors who, by 1837, had provided him with specimens. Of this latter group, he ob serves, few "are sufficiently skilled in pinning and preparing" insects.[76] The basic categories of information exchange are provider and recipient and initially it was hypothesized that, during the course of Harris's career and with an increasing personal expertise, there would be a movement from one who received to one who provided information. The content of a number of his letters was analyzed, noting statements that indicated that he was provider or recipient of information. Looking at these statements in the aggregate, his role was more-or-less equally provider and recipient over the course of his career, and when viewed at regular intervals over the years, there was fluctuation but little discernible pattern. It was only when the statements from the letters were looked at for two personally significant periods that any noticeable difference was apparent. For the years when he was most active as a researcher in entomology, the decade from 1833 to 1842, the data from the analyzed letters indicate that he was notably less likely to be a provider and more likely to be a recipient of information. For the years after 1843 the roles were reversed, and he was provider more often than recipient.[77] The tentative suggestion from this result is that when a naturalist was deeply involved in investigation, the likelihood (necessity) of receipt of information from the scientific community was greater. After 1842 and the decline of his entomological career and therefore of his own information needs, Harris became a community resource and the instances where he was provider of information to others increased. Whether the receipt of information through correspondence was equally important for other Americans actively engaged in natural history research at the time, for Europeans presumably having access to more established collections and library resources, or for naturalists in general for later generations cannot be answered by these tentative conclusions from a functional analysis of Harris's letters.[78] What they suggest for Harris is the deep mutuality of his involvement in the scientific community, the importance of community itself, and the consequences for his role in that community from the events in 1842 and the diversion from natural history that his duties as librarian forced on him thereafter. These are facets of his life and career affirmed through even a general acquaintance with his life course and from a familiarity with the substantive content of his letters.

RELATIONSHIPS WITH INDIVIDUALS

The individuals with whom Harris interacted in the conduct of his entomological and other natural history work were numerous and varied in character. Based on his correspondence at hand and his acknowledgments in his publications,[79] about 145 people have been identified as involved directly as providers or recipients of information (broadly defined). Some seventeen were European naturalists and he summarized these contacts in 1836 for his American friend David H. Storer:

> I have . . . received some very kind & flattering letters from foreign naturalists, among whom are Dr. Hooker the botanist of Edinburgh, who did me the honor to present to me a portion of Lyonet's posthumous works, after reading the remarks which closed my catalogue of insects in Hitchcock's Report. The notice of this gentleman was as unexpected as it was grateful, and was owing to the interest which his son feels in the study of entomology. The son has also sent me some fine insects, & promises more in return for such as I may send to him. The well known naturalist Wm. Sharpe Macleay & Mr. J. O. Westwood, Secretary of the London Entomological Society have also honored me with letters; and the former promises to aid me in my favorite studies "in any way which I may be pleased to point out"; and the latter has sent me an interesting & valuable suite of the English bees. Thro' these gentlemen & from Pecchioli, librarian of the Grand duke of Florence I hope to get collections of nocturnal lepidoptera & diptera to illustrate the modern genera in these orders, which require to be studied in this way.[80]

In spite of his critical approach to the practices of Europeans vis-à-vis American species, it is apparent, therefore, that these relationships, in addition to bringing material help, also were an important part of Harris's sense of involvement and acceptance in the community of science.

About forty percent of the Americans involved in Harris's network are not recognized as naturalists in the most comprehensive listing for this time period.[81] They included individuals thanked in Harris's publications for the insects and information they had provided. For example, he writes, "My attention was called to the depredations of this bud-moth, and of the preceding species, by John Owen, Esq., of Cambridge, by whom the moths were raised from the caterpillars, and presented to me."[82] Dr. Ovid Plumb of Salisbury, Connecticut, warranted reference in the revised edition of Harris's *Treatise* for information and specimens he provided, including Plumb's experiments on plant lice and possible remedies for the insect's damages. In July 1851, Harris traveled to Salisbury, where he made observations of his own.[83] Several students at Harvard are credited (either by name or anonymously) for specimens provided to Harris, from Cambridge and from as far away as Maryland and Virginia.[84]

Harris got considerable help with his investigation of the wheat-ly through letters received during 1838 and 1841 from Mrs. Nancy George Sibley Gage, of Hopkinton and Concord, New Hampshire, who was married to Dr. Charles P. Gage. She conducted what appar-ently were extensive observations of the insects and provided Harris with reports as well as specimens.[85] Mrs. Gage was the cousin of John Langdon Sibley who worked with Harris in the library, and initially her letters went to Sibley who transmitted them to Harris. In one she wrote, "Father says too, success to Dr. 'H. in his researches; & if he can ascertain a practicable & profitable way of exterminating these destroyers, he will confer a blessing on his country & the world not less in value than Davy did in his safety lamp['?]."[86] Gage's father, Stephen Sibley, Esq., of Hopkinton, provided information on another insect, which was acknowledged by Harris in the first edi-tion of the *Treatise*.[87] By 1841, Gage was writing directly to Harris in regard to her entomological interests and investigations.[88]

Harris engaged in entomological and related correspondence with individuals not referred to in his publications, including a fair number who are not known as naturalists.[89] There were those who wrote to him seeking information on insects that were causing ag-ricultural damage. Harris's early contacts with non-naturalists also included individuals who could provide insect specimens for his de-scriptive and taxonomic work, including especially Charles J. Ward of Roscoe, Ohio, with whom he corresponded in the late 1830s.[90] Edwin Willcox, resident in New York City, also was an entomological correspondent in this time period. As noted earlier, Willcox bought John Kirk Townsend's Oregon insects that had been placed in Har-ris's hands by the Academy of Natural Sciences of Philadelphia and it was this event that brought an unhappy end to their relationship. Though Harris and Willcox had had a constructive relationship pre-viously, this incident led Harris to pronounce Willcox a "mere trader in insects, and not susceptible to any of those honorable feel-ings which should actuate the lover of science."[91]

Harris's correspondence with non-naturalists in the period after publication of his *Treatise* reflects the reputation that the work earned him. Dr. John M. B. Harden wrote him from Liberty County, Georgia, in 1844 to ask for advice on an insect he had found.[92] His relations with Thomas Affleck, agriculturist and planter in Missis-sippi, have been referred to earlier. Affleck's 1846 proposals for col-laboration on the cottonworm (Harris to do the technical description and Affleck to contribute his observations of the insect in the field and its damages)[93] illustrates how the informed and ex-perienced naturalist and the interested layperson could work together to a common goal. Affleck was, of course, more than a curious inquirer, and the fact that he does not appear among the naturalists listed in Meisel's *Bibliography of American Natural History* is a measure of the way in which agricultural and natural history interests in the period were normally disconnected.

In November 1848, Harris was contacted by Stephen Calverley of New York, who reminded him that they had been in contact several years earlier and that Harris had then agreed to name some insects for him. Calverley's wife had become ill and he had to discontinue work on his collection, but was now ready to resume its arrangement. He also inquired whether Harris's state report on injurious insects was available in the bookstores, which suggests that the inquiry in this case was based on Harris's general entomological reputation rather than derived from that book in particular.[94] Harris responded that his duties in the library prevented him from giving quick attention to the task of naming insects, but Calverley was happy for any assistance he could get. Harris did send a copy of his *Treatise* (for which Calverley sent him two dollars). In response to an inquiry from Harris as to how Calverley had come to name his butterflies, he stated: "a list of Butterflys found in the U.S. which I understood you furnished to Mr. Hooper, some time ago, was given to me he having had a Lithograph copy made, which served me as a guide to species, I have works containing figures of most of the Butterflys, with the exception of some few small species which I will send to you, Mr. Doubleday gave me the names of many of them on his return from your place when he was in the U.S. and Major Le-Conte has given me the names of some of them."[95] This is the only reference to a list of butterflies prepared by Harris, and Hooper is not otherwise known.

In 1851, Harris became involved in a controversy that began in a private exchange but subsequently became public. The incident suggests that not all in the community were as deferential to him in these matters as John W. Proctor, of Danvers, Massachusetts, with whom the exchange on the potato worm seems to have begun. Proctor observed the worm on the vine and wrote to Harris in July with the semifacetious observation that "I know of no place where they [the insects] can be subjected to a trial so perfect as before yourself; your treatise 'on insects injurious to vegetation' having established your right to be considered Chief Justice in these matters."[96] Proctor published Harris's reply in the *Salem Observer,* and a Mr. Whipple apparently took exception to the views expressed there. A brief public exchange resulted, but finally Harris concluded not to continue the discussion, noting that "In my remarks on Mr. Whipple's interesting

observations, I endeavored to express my own views in such respectful language as should at least give no offense."[97] Although the details of this event are not entirely clear, it does suggest the character
of the controversy that could develop around insects when agricultural interests were involved, and the ways in which expert knowledge (Harris) still had to contend with an array of other voices that
could draw on tradition, personal observation, and perhaps other
forms of social authority.[98]

Within the American naturalist community, Harris had a number
of ongoing relationships that were mutually beneficial, although the
particular individuals changed over time. As discussed in a previous
chapter, his first important relationship was with William Dandridge
Peck, professor of natural history in Harvard College who became
Harris's mentor and friend. In the early 1820s, he contacted
Thomas Say and carried on a correspondence with him and was able
to get Say to examine and name a number of his insect specimens.[99]
Harris's responsibility and anguish over Say's own collection, which
came to him after his death, has been discussed above. On a more
personal level, Nicholas Marcellus Hentz and Charles Pickering, especially the former, were Harris's particular friends in entomology
in the 1820s. The origins of Harris's friendship with Hentz are not
determined, but they were in contact by correspondence as early as
1824. Hentz had resided in Boston while teaching French and min-
iature painting and it is likely that they met then. Their correspondence in the 1820s shows a mutual support for their natural history
interests as both men struggled to find time, resources, and an outlet for their ideas and work. Hentz exclaimed in 1826, "Oh! Why
must we live at such distance from each other! What pleasures we
might enjoy together! I feel the want of books still more than you
do."[100] Later that year, Harris visited his friend in Northampton
where the two men seem to have thoroughly indulged their naturalist interests. Afterwards he wrote, "Please say to Mrs. H- that I regret
the shortness of my last visit to you, & the absorbing nature of our
studies should have prevented me from more intimately enjoying
her society & cultivating her acquaintance . . ."; "As to yourself I can
only say that I shall reckon the time passed with you as among the
brightest moments of my existence." The letter continued, several
weeks later, by which time Harris had learned of a "domestic calam-
ity" in Hentz's life, apparently the death of a child at birth. Harris
noted illness in his own family, but in keeping with the scientific
bond that supported their friendship also mentioned that he would
send Hentz the last two volumes of Kirby and Spence's *Introduction
to Entomology*.[101] By the time this letter was written, Hentz was prepar-

ing to leave for the South, where he spent the remainder of his life, teaching at the University of North Carolina and at private schools for girls that he operated with his wife. Although Hentz's removal to the South deprived Harris of a companion in the study of the insects of Massachusetts, it perhaps helped to expand his geographical perspective on entomology. For a time the exchange between the two men continued in intensity and, in 1828, Harris suggested that their letter writing be regularized, Harris to write the middle of each month and Hentz the beginning of the next.[102] In 1830, Hentz named a son for his Cambridge friend.[103] After that time, however, their exchanges appear to have been much less frequent.[104] By the mid-1830s, Hentz had largely abandoned active study in entomology[105] and, as discussed in an earlier chapter, Harris devoted considerable effort in the procurement of his collection for the Boston Society of Natural History. Hentz is most noted for his studies of spiders, and the last letter that has been seen between the two men (1845) concerned Hentz's shipment of a manuscript on the Araneids to Harris, noting, "I commit to you the work of twenty three years."[106]

In an era when serious students of natural history, and entomology in particular, were few, finding a new one was an event similar to finding a new species. Charles Pickering was only twenty years old in 1826, and ten years younger than Harris, but the older man recognized his abilities and took note of his advice. Writing to Hentz, Harris reported that "I have lately become acquainted with Dr. Charles Pickering of Salem [Massachusetts], who is now attending his 3d course of med. lect. in Boston. He is a young man ardently devoted to nat. sc., has been a pupil of Mr. Nuttall,[107] & is an excellent botanist, & good entomologist. He urges me to study the moths (Lepidoptera nocturna Latr.)."[108] Pickering offered Harris advice and crucial assistance in his entomological work, helping, for example, to arrange contacts with European entomologists for exchange.[109] He also appears to have played the role of intermediary in Harris's relations with the Philadelphia naturalists,[110] including the arrangement with the Academy of Natural Sciences to send Say's collection to Harris.[111]

The bulk of available letters exchanged with Pickering do not extend beyond 1838,[112] but Harris did advise Pickering in regard to suitable naturalists to accompany the U.S. Exploring Expedition to the Pacific (Wilkes Expedition) of 1838–42, for which Pickering was chief zoologist. In October 1836, when preparations for the Expedition were getting underway, Harris wrote: "It gives me pleasure to find that measures are taking [sic] to engage an entomologist for the expedition . . . Mr. Herrick of New Haven is well known to me, and in many respects would be a good person. There are others, however, who would probably do more than he could, & who have enjoyed greater advantages to prepare themselves for the service. Among these first, decidedly, is Dr. Augustus A. Gould of Boston. He has received a University[113] & medical education, & is competent to the charge of several branches of Natural History. For two years past he has performed the duty of Lecturer on Botany & instructor in Zoology to the Senior class of our University, & will probably continue to discharge that duty until a professor is appointed." Harris, who would himself take over these teaching duties from Gould the next year, pointed out that his colleague was particularly knowledgeable in invertebrate zoology, noting that he is "the best of our conchologists," and "as an entomologist I know him to be full of zeal, & sufficiently acquainted with the science to give it proper attention." Harris also suggested Abraham Halsey, of Hartford, Connecticut, whom he knew only through correspondence. Further noting that, while Herrick's abilities might be developed, he considered him "inferior in qualifications to Dr. Gould, & in practical skill to Mr. Halsey. With the exception of these gentlemen I know of no one who would do better than Mr. Herrick."[114]

In the same month, Pickering inquired about John Witt Randall and his suitability to serve as entomologist on the Expedition. Harris replied that "Mr. Randall has been known to me about six years. He first sought my acquaintance when he was freshman in the University. I found that he had some taste for natural history particularly Botany & Entomology." In evaluating Randall's development, Harris observed that "I cannot promise that he will ever become a thorough Entomologist, but I am well persuaded that he would be an industrious collector, and as such would be a valuable acquisition to the corps of savans."[115] Shortly thereafter, Harris wrote again to Pickering to inquire about Randall's prospects for the Expedition and reported that "He is now applying himself closely to the study of genera & species under my direction."[116] In the end, however, Randall was not included in the Expedition, an event that appears to have affected him considerably. In 1838, Harris wrote that, Randall, having been cut from the Expedition, was inclined to complete his medical studies with his father, something which "all his friends wished & urged him to do some years ago."[117] Harris had continuing hopes for Randall as an entomologist, but wrote the following year that Randall was "so much mortified & disappointed" at not having accompanied the Wilkes Expedition "that he has lost all ambition to [pursue the?] natural history[.] I have endeavored, but, in vain to rouse him to exertion—he never would bring himself down in earnest to the technicalities of Entomology & is now even unwilling to attempt to describe even some of our new species."[118] Whatever Harris thought of other candidates for the Wilkes Expedition, as the excursion got underway, his confidence in Pickering was unwavering, describing him as "the soul of the corps of naturalists embarked in the South Sea expedition."[119]

Although not much is known in detail about Harris's relations with William LeBaron (1814–76), it is clear that he had an important influence on this leading agricultural entomologist of the subsequent generation. A medical practitioner, LeBaron was living in North Andover, Massachusetts, when he opened a correspondence with Harris in 1838. As such, he was able not only to write to Harris but to visit with him as well, and in 1841 he described himself "in some degree as your entomological pupil." In 1844, LeBaron moved to Illinois and later served there as state entomologist.[120]

Harris described his classmate Levi Leonard in 1828 as a "new recruit in Entomology." Settled as a Unitarian clergyman in New Hampshire, Leonard was attracted to the study of insects through reading Harris's articles in the agricultural press.[121] Harris wrote to him and encouraged him in his entomological contributions: "I assure you that, through your zeal & aid, much useful matter may be added to our very imperfect Entomology. I may possibly tax these, as well as your patience, to their full extent in the progress of my enquiries into the habits & oeconomies of insects."[122] Harris in turn assisted his friend through the loan of books.[123] Although Leonard became a serious student of entomology, he does not appear to have had any particular ambitions for himself. In fact he seems to have adopted the role of information provider for American science—he served 1849–52 as an observer–contributor for the Smithsonian Institution's meteorological project.[124] Thus, while Leonard more properly was a non-naturalist associate of Harris—among the group discussed above who supported his work without undertaking publication in their own right—through his persistent study of entomology he was much more important to Harris than most of the others. The magnitude of his contribution is indicated by a report to Hentz in 1828 that he had received 500 to 600 Coleoptera from Leonard, "and many of them are new to my collection."[125] Leonard also was described as "indefatigable in studying the habits of insects," and Harris noted that his information on the larvae of many insects in his collection, especially the Lepidoptera, was received from his friend.[126] In fact Leonard was put in a category other than collector. Harris referred to him in 1839 as among "one or two accurate obervers," "who, if he had time & health to attend to the subject would make one of the best practical entomologists."[127]

In the *Treatise*, Leonard's assistance was acknowledged more frequently than any other individual, with help in several insect orders noted. In discussing the skippers among the Lepidoptera, Harris wrote, "For a specimen of the male I am indebted to the Rev. L. W. Leonard, to whom I have dedicated the species" as *Hesperia Leonardus*.[128] Naming an insect for Leonard was intended as a special honor and recognition, but on the surface it might seem a curious contradiction, since Harris professed an "insuperable objection to having any insects of the United States bear my name as a specific appellation." So strongly was Harris opposed to having an insect named for him, that he asked J. E. LeConte to choose another one.[129] But from other evidence, what may seem like modesty on Harris's part really works to separate him as the committed and publishing naturalist from Leonard the aide and collector. So when the European entomologist C. J. Schonherr proposed naming an insect for him, Harris wrote to a colleague, "I am glad that Schonherr's Brachystilus does not prove to be new; for I do not want to have my surname [*sic*] tacked to any American insects; because I expect to describe these insects myself, & it would to me be very unpleasant to describe any with 'Harrisii' affixed as a specific appellation."[130]

Harris's other early and important naturalist contacts included John Eatton LeConte and Edward C. Herrick, both of whom have been referred to at several points above. It was through his father, Thaddeus Mason Harris, who traveled to Georgia in 1834, that Harris came into contact with John Abbot, whose illustrations and notes had been foundational to James E. Smith and Abbot, *The Natural History of the Rarer Lepidopterous Insects of Georgia* (London, 1797).[131] Abbot wrote to Harris, sending him several drawings and a box of insects,[132] and Harris responded with a query regarding the price that Abbot charged for his drawings and promised to send him some insects.[133] No further direct contact between the two men is noted, but when Harris began to plan in 1838 for his publication on the Lepidoptera under the auspices of the American Academy of Arts and Sciences, he expected to have Abbot's assistance with the illustrations.[134] In 1839, however, he asked LeConte, for whom the artist had worked, whether Abbot still was living and able to render help, which tends to confirm the lack of any continuing contact between Harris and Abbot.[135] Abbot was known not only for his work as an illustrator but also for his study of the life histories of insects,[136] and Harris considered him his only predecessor in the study of the American nocturnal Lepidoptera.[137] Smith and Abbot's *Natural His-*

tory was an important and acknowledged source of descriptions of the caterpillars in Harris's 1839 catalogue of the Sphinges.[138] Some years later, Harris reported, "A young Dr. Oemler of Savannah, son of a deceased correspondent of mine, has lately sent to me for examination a collection of drawings, made by old Mr. John Abbot of Georgia, one hundred and sixty-three in number, representing butterflies, sphinges and moths of that State, with the plants on which they feed. They are not correctly done, but in general are enough like the original to be recognized, with perhaps a few exceptions only."[139] Most likely Harris had these illustrations in mind when he approached Joseph Henry at the Smithsonian Institution. In February 1855, assistant secretary of the Smithsonian Spencer Baird wrote to Harris to convey Henry's reaction to the idea of publishing a supplement to John Abbot's Lepidoptera of Georgia. In declining the project, Baird reported that the Smithsonian had a practice of not taking on publishing ventures that had been started elsewhere. Baird, however, did make a counter proposal, thinking that the same purpose could be accomplished if Harris prepared a "manual of genera & species of N. American Lepidoptera with these figures as illustrations." He thought that the Smithsonian would be able to publish a work in this form, should funds become available.[140] There is no evidence, however, that Harris ever followed through on this proposition.

Harris also had a sustained correspondence with Pennsylvania entomologist Frederick E. Melsheimer, whose father, Friedrich Valentine Melsheimer (1749–1814), had published *A Catalogue of Insects of Pennsylvania* (Hanover, PA, 1806), the first separate publication on American insects (although it included only beetles).[141] The correspondence began in 1835 when Harris sought help with Say's insects and republication of Say's New Harmony papers on Coleoptera[142] and continued at least until 1844. Melsheimer was one of Harris's important contacts in support of his special entomological interests—Harris wrote in 1839, "It gives me great pleasure to find that you have paid considerable attention to the Lepidoptera, & particularly to their larvae & habits, which are essential to the proper arrangement of these insects."[143] A decade later, however, he reported to Doubleday on the limits of such long-term and long-distance relationships, noting that "Dr. Melsheimer, who is . . . a good observer, lives far from me, retired in the midst of the woods of Pennsylvania."[144]

Harris's relationship with the somewhat enigmatic Charles Zimmerman[145] was closer than with some other entomological acquaintances. Although Zimmerman is not mentioned in Meisel's

Bibliography, he published several papers in German journals between 1832 and 1843, and he was well known among European entomologists prior to moving to the United States.[146] In his relations with Harris and the American entomological community he proved to be a serious naturalist and collector, and was respected for his taxonomic knowledge and insight. Harris's relations with Zimmerman have been alluded to in previous chapters, especially his plea to Zimmerman early in 1841 to come to Cambridge and help with the Say Collection when its return had been unexpectedly requested by the Academy of Natural Sciences of Philadelphia.

In writing of Cambridge visitors who came to dinner at the Harris home, Harris's daughter wrote, "Dr. Zimmerman, the entomologist, was one . . . whose repeated visits made him a loved and familiar friend with the children, as well as with the parents."[147] Zimmerman's own fond remembrance of the Harris children has been referred to in an earlier chapter. From the available correspondence, however, the only confirmed time that Zimmerman was in Cambridge was the fall of 1839. He was residing in Columbia, South Carolina, and somewhat whimsically, it seems, wrote to Harris in May 1839, "I think to proceed northward, and to fix myself somewhere near Philadelphia or Baltimore. There I intend to become a citizen, to build a house, to buy some lands, to plant corn, feed silkworms, catch new insects, study your entomological writings, and ask your personal visits."[148] Apparently, instead, Zimmerman decided to return to Germany, a decision that led to disaster for an entomologist, when his insect collection and books were lost at sea.[149] Harris reacted with concern, understanding, and charity: "Most truly do I grieve & sympathize with you for the loss that you have sustained," and proposed a plan that would resurrect the resources of the suffering entomologist: "first you must lay all of us under contribution; I doubt not that [all?] your entomological friends in this country will cheerfully contribute to the extent of their ability. In my own cabinets there are still many duplicates, both native & foreign insects. If you will take the trouble to look them over & make a selection from them you are welcome to all that you want. Moreover, you know that I have a claim upon the duplicates in Say's collection—and from these, as soon as the selection for the Academy is secured, you shall have a pick." He also offered "a suite" from the collections of the Boston Society of Natural History. In order to carry out this plan, he invited Zimmerman to spend the winter in Cambridge, "where you may live at from 5 to 8 dollars per week." Harris added, "In the Spring, if you please, you can make excursions into Maine, New Hampshire, & Vermont; and I think that you can easily persuade

~~John~~ [*sic*] young Randall to accompany you, if you wish a companion." He also suggested that Zimmerman might want to consider going West with a scientific expedition.[150]

Shortly afterward, in early October, Zimmerman apparently was with Harris, since Harris wrote to his friend William Oakes at Ipswich about arranging a trip there with Zimmerman to collect water beetles.[151] Harris obviously gave some time to Zimmerman, because in November he wrote to Melsheimer, "My numerous engagements & the attention which it became necessary for me to bestow upon our friend during his residence in Cambridge, have prevented me from finishing the examination of the insects which you sent to me." By this time Zimmerman had left Cambridge, but Harris expressed his hope to Melsheimer that Zimmerman would take up permanent residence there.

> In this case much more could be done by us together in promoting the cause of Entomology than by both of us separately. The Dr. will find here collections & books, which he could not meet with in any other place in the union. They would be of immense advantage to him in completing his system of Coleoptera, which it is very desirable should be published. The Dr. is one of the most accurate observers, and, being near-sighted, possesses a great natural advantage in examining the minute characters of insects. He has given me reason to think that he will return and live here, and I should regret, on my own account very much & still more on account of science, if anything should happen to alter his present intentions.[152]

By the time Harris wrote to Zimmerman in February 1841 for help with the Say collection, he had lost track of his friend's whereabouts and wrote to him by way of John G. Morris who resided at Baltimore. When Zimmerman responded, he was back in Columbia, South Carolina, reporting that when he left Harris in November 1839 he had gone to New York and then to Baltimore and remained there and in the vicinity until the summer of 1840. He announced that he was about to move to Rockingham, North Carolina.[153] The two men apparently never met again, although their correspondence continued. In 1846, Harris received a most curious letter from Zimmerman, who was responding to a package that Harris had sent. "Among the rest [of the contents] I found to my surprise more than I expected, viz. a young eagle imprisoned in a paste board box. I have kept it there all the while, without food, to be carried back again. This bird came among the beetles, I suppose, as Saul among the prophets—by mistake, and if so, it is evident that mistakes may be committed even at Harvard University, to say nothing of the rest of the world, which, everybody knows, is chokefull of them."[154] Although Zimmerman's letters tended to be somewhat facetious and politically flavored, there is no way to know whether it was true of this letter or whether he, in fact, was entirely serious. The last letter that has been seen from Zimmerman was written in 1853, and he was still at Columbia.[155]

During the exhilarating years in the late 1830s and early 1840s, when Harris's entomological work crested, the English Quaker entomologist Edward Doubleday was his most important scientific associate. Their relationship has been referred to in a number of contexts above. Not only did they develop a deep friendship in science, but through their correspondence they were able to share the trials and highpoints of their personal lives as well. A son born to Harris in 1839 was named for Doubleday. The English entomologist was born in 1811 and therefore some fifteen years younger than Harris. Perhaps because of Doubleday's situation as an English naturalist and, after 1839, as assistant in the British Museum in charge of moths and butterflies, however, Harris seems to have been the primary beneficiary from their professional relationship, Doubleday's location in the metropolis of science and his institutional situation overcoming what might have been a more natural factor of age that would have given Harris the advantage. Late in their relationship, Doubleday reported on his work load, noting that "The Museum occupies me six hours a day, and I live three miles from it, so that is eight hours occupation daily. My book takes me full thirty-six hours a week, or five hours daily. I belong to eight London societies, and am in committees or on councils of several. I have to do a great deal to help Boisduval and Guénée in the Suites à Buffon, and occasionally lend a hand in other books. I have correspondents from Labrador to Valparaiso, from Copenhagen to New Zealand. You will see that this is plenty of work."[156] Though Harris could commiserate with his friend in regard to the work that administration of an institution entailed, he, nonetheless, envied Doubleday's situation. "You have no idea how much I miss the advantages to be derived from books, from cabinets, & from an interchange of opinion with skilful entomologists, all of which you enjoy. None of our entomologists know anything about Lepidoptera scientifically."[157] Some months after this letter, on December 14, 1849, Doubleday died at London. With his death Harris lost even that link to the world of research entomology, and to his one-time hope to make a substantial contribution to the study of the Lepidoptera.[158]

In the period after 1842 and especially in the late 1840s and the 1850s, Asa Fitch and John Lawrence LeConte were among Harris's particularly constant entomological correspondents, relationships noted in the previous chapter, as was his correspondence with and about Margaretta Hare Morris in regard to work on the Hessian fly. Among the other women with whom Harris corresponded on entomological matters was Nancy Gage, mentioned above, who also studied the Hessian fly and related insects. In his 1835 catalogue of insects of Massachusetts in Edward Hitchcock's state report, he thanks Miss D. Dix "for several interesting insects."[159] This was, in fact, Harris's cousin, Dorothea Lynde Dix (1802–87), the humanitarian reformer known for her work on behalf of the insane. Harris had an interesting relationship with Dix, although next to nothing of it has been seen in his own papers. Two recent biographies of Dix give a picture of her relations with the Harris family. After Dix moved to Boston about 1821 to live with her grandmother, she was strongly influenced by her uncle (and Harris's father), Rev. Thaddeus Mason Harris, whose Dorchester church she sometimes attended. In this situation, she came under the influence of Thaddeus William who is described as a mentor or tutor to Dix in nature studies, and she began to collect on her own. In December 1825 she spent the month with Thaddeus William and his family in their Milton home, and it apparently was he who suggested several years later that she contact Benjamin Silliman, professor of chemistry and natural history at Yale and editor of the *American Journal of Science*, about her scientific interests. This step initiated an exchange of letters between Dix and Silliman that continued for more than three decades. Dix attempted to prepare a botanical paper for Silliman, but then turned to insects. A brief and somewhat eclectic paper on the latter subject appeared in 1830,[160] and, though related to Harris's interests, it made no reference to him. It appears that Dix had an ambivalent relation with Harris as she did overall with her family. Harris is described by one biographer as her favorite cousin, while another concluded that, as of the late 1840s, "she cordially despised her most prominent Boston cousin." They do appear to have kept in touch, however, and reference is given to her transmittal of a box of insects from Frederick Melsheimer (who resided in Pennsylvania) to Harris in 1840.[161]

Harris's relationship with Henry David Thoreau was explored some years ago by Joe Sanford Wade, a scientist at the U.S. Department of Agriculture's Bureau of Entomology.[162] Although Wade's published account contained a number of errors, his quoted passages from Thoreau's journals regarding his visits to Harris and comments on the Massachusetts' reports on natural history are genuine and offer an interesting glimpse of how Harris was observed by a naturalist with a different orientation from his own. Wade's selections from Thoreau's journals document visits that occurred in the years 1852 to 1855, all of which probably took place in the library. Thoreau also wrote at least one semipersonal letter to Harris, in 1854, by which he returned books to the library but also included an insect specimen collected in Concord. He commented on the insect's sound, which he reported to have heard on specified dates in years preceding.[163] In Wade's assessment of Thoreau's notebook entries, Harris is characterized as the teacher to the younger naturalist[164] as he identified specimens that Thoreau brought to him.[165] Some entries in the notebooks also relate to Harris's historical interests. For example, Thoreau learned, from a visit on February 9, 1853, about Harris's views on the early history of the potato, a topic that Harris was investigating at the time.[166] Harris, in fact, was Thoreau's early teacher in natural history, since it was he who taught natural history in Harvard College when Thoreau took the course as a senior in spring of 1837.[167] We do not know of their personal relations when Thoreau was an undergraduate, but the young student began his lifelong use of the College library during that early period,[168] and he and Harris undoubtedly encountered one another there as well as in the classroom. Thoreau's friend and earliest biographer, in fact, took note of the long-term double relationship that Thoreau and Harris shared, revolving around both the library and natural history, commenting "Many are the entries in his Journal of his visits to Cambridge where he went to get his books from the College Library and to have a chat with his valued friend, the naturalist Dr. T. W. Harris."[169]

Thoreau reviewed the reports of the Massachusetts survey in 1842, in his essay on the "Natural History of Massachusetts,"[170] although, as John Hildebidle notes, the paper did not consider the content of the survey volumes very specifically at all.[171] The review reflects Thoreau's ambiguities about nature and scientific study, positing himself, as one interpreter has noted, "as the literary competitor of the men who wrote the Massachusetts Reports—as an antagonist, in fact, who needed to establish his own authority." He opposed the view that the naturalist is an outside observer of nature.[172] Thoreau commented in the "Natural History of Massachusetts" that entomology "extends the limits of being in a new direction, so that I walk in nature with a sense of greater space and freedom. It suggests besides, that the universe is not rough-hewn, but perfect in its details. Nature will bear the closest inspection . . . She has no interstices . . . ,"[173] thus interjecting how the study relates both to himself and to what it reveals about the natural world. But there are no specific references to Harris's report aside from the observation that, along with the reports on fishes, reptiles, and invertebrates, it shows "labor and research, and . . . [has] a value independent of the object of the legislature."[174] It is not clear that Thoreau actually had Harris's report in hand for this review. Although he acquired most of the reports, he apparently did not own Harris's work until the appearance of the second edition of the *Treatise* in 1852.[175] Publication of this new edition would have coincided in time with some of the extracts Wade took from Thoreau's journal and which indicated an interest in insect studies. It is possible, therefore, that Thoreau sought out Harris at that time as a result of his ownership of the *Treatise* and the close reading that possession would have permitted him. During this time period, Thoreau, though with mixed sentiments, found himself turning more to a detailed study of nature, buying books and traveling to the headquarters of the Boston Society of Natural History (and to Harris in the Harvard library), acts that brought him a sense of involvement with a community of study.[176]

While Thoreau was able to interact and benefit from his relations with Harris as his mentor in entomology, he turned the material to his own purposes while he questioned the social or political motives that had supported Harris's chief work in science. In a passage in his journal in 1852, he noted that Harris had identified a "great moth" he had brought to him as the *Attacus luna,* seldom seen because it was eaten by birds. Harris told him that "Once, as he was crossing the College yard, he saw the wings of one coming down, which reached the ground just at his feet," and from this Thoreau drew the metaphor that the work of the poet was like such wings, the only evidence of their author's soaring flight.[177] After Harris's death, Thoreau still was critically contemplating the basis of public interest in entomology, writing in his journal in 1859 how the state chose to give attention only to the bad or threatening in nature :

> This occurred to me yesterday as I sat in the woods admiring the beauty of the blue butterfly. We are not chiefly interested in birds and insects, for example, as they are ornamental to the earth and cheering to man . . . the only account of the insects which the State encourages is of the "Insects Injurious to Vegetation." We too admit both a good and bad spirit, but we worship chiefly the bad spirit, whom we fear. We do not think first of the good but of the harm things will do us.
>
> The catechism says that the chief end of man is to glorify God and enjoy him forever, which of course is applicable mainly to God as seen in His works. Yet the only account of its beautiful insects—butterflies, etc.—which God has made and set before us which the State ever thinks

of spending any money on is the account of those which are injurious to vegetation! . . . Come out here and behold a thousand painted butterflies and other beautiful insects which people the air, then go into the libraries and see what kind of prayer and glorification of God is there recorded. Massachusetts has published her report on "Insects Injurious to Vegetation," and our neighbor the "Noxious Insects of New York." We have attended to the evil and said nothing about the good.[178]

It is unlikely that any such sentiment was derivative from Harris himself. Devoted to his science and interested in the full scope of entomological investigation, Harris was open to an opportunity to publish the results of his studies and to accept the prescriptions of the state legislature. That he never was able to prosecute his more technical studies in entomology, and especially of the Lepidoptera, Harris would have attributed to the lack of an academic opportunity rather than wrong-headedness on the part of the Commonwealth. For his part, Harris showed how friends can differ when he commented that "Thoreau would be a splendid entomologist if he had not been spoiled by Emerson."[179]

In certain respects, Louis Agassiz represented the antithesis of the kind of natural history that Thoreau sought to practice. Thoreau abhorred the classificatory and museum-based science of Agassiz and his personification of the process of professionalization that moved others ever further out to the margins of nature studies. In fact, Thoreau's characteristic concern with "the living relationships of animals and plants to each other, to himself, and to mankind"[180] is a likely source of attraction to Harris, whom he might have seen as a naturalist outside the constraining community of professionals that Agassiz epitomized. Harris did have a relationship of mutual respect with Agassiz, however, and some of the incidents of their interaction have been noted at several points earlier, including Harris's work on Agassiz's Lake Superior survey Lepidoptera and their exchange about Agassiz's *Nomenclator Zoologicus* and the general topic of names in natural history. Harris consulted Agassiz on entomological matters and also was able to benefit from the availability of publications in Agassiz's possession. However, his situation limited the degree to which he could make use of Agassiz's advice and resources, writing to Edward Doubleday in 1849, "With Professor Agassiz I am well pleased, & he has ever treated me with most respectful consideration; but my duties have not permitted me to see as much of him & of his doings as I could wish, or to avail myself of his many kind offers of the use of his books & specimens."[181]

In an 1848 letter to Agassiz,[182] Harris explained at length his views

on the functions of the veins or nervures of insects, following up on
a conversation that the two men had had "on Tuesday morning . . .
by way of apology for my caution relative to the function of the wing
veins or reticulations."[183] Harris explained,

> It is highly desirable that the structure of insect's wings should be thor-
> oughly examined, by means of high magnifying powers, in the hands of
> persons capable of using these instruments; and I have had neither in-
> struments nor time to do this myself thoroughly. In consequence, it may
> be that my notions on the subject are erroneous. It would be a curious &
> interesting subject of study to ascertain fully the structure & uses of the
> veins, as they are called in common language; to learn whether any mus-
> cles & nerves pervade or accompany these parts, & how or by what means
> the wings are folded longitudinally & transversely, as the case may be, in
> various insects.

He asked further, "Are the wings aerial organs in any other sense
than as instruments for flight? It has been my opinion that they
are—that they are pervaded by tracheae, whose office is to convey
air only: That they are, as it were, external lungs. At the same time,
I admit that they contain fluids." Harris went on to explain his dis-
agreement or skepticism over certain conclusions that the veins
functioned to circulate fluids, arguing that it is known that they con-
veyed air in some insects and therefore assumed a similar function
in all insects. He argued, furthermore, that in addition to the conduct of air, the veins also served a structural function in the wing.

To explain his idea about how fluid circulated, he made a botanical comparison: "The wing of an insect, like the leaf of a plant, may
be likened to a kind of bag, laterally compressed, so as to have its
interior surface brought into close contact throughout . . . between
which pass the distending tubular veins & their reticulations. All
along the sides of the veins there will therefore be longitudinal cavit-
ies, serving as channels for the passage of fluids." In this way, in Harris's view, the same basic structure was able to allow for both air
passage and the flow of fluids, while keeping the two separated. He
suggested that it would be an interesting undertaking to compare
the vessels in insect wings and in plants.[184] Harris did not manifest
any great interest in insect anatomy, in general, and some years before he had written that, while it was a subject of study in France, "I
confess the subject is not to my taste."[185] (Perhaps the French connection itself was among the reasons for his lack of interest.) It is
clear from the letter to Agassiz, however, that he was able to generate a keen interest in such structural and functional questions when
the occasion arose, and perhaps the opportunity to carry on such a

discussion with Agassiz was itself an incentive. Interestingly, in this letter there is no reference to the use of wing veins as an aid to classification, a topic of considerable interest to him at one time, as discussed in the previous chapter.[186]

It is difficult to reconstruct the relationship that Harris had with Asa Gray. Certainly Harris harbored considerable resentment about the loss of the Fisher professorship of natural history to Gray, but the degree to which that extended to the man himself is not certain. The general lack of surviving correspondence between the two men may be a gauge of the small interaction they had, or it may only mean that their relationship was one of close proximity at Harvard. However, in the early 1850s, when Harris was working on the squashes and pumpkins, he sent an "elaborate paper" on the subject to Agassiz and apparently not to Gray, to which Agassiz replied, "No doubt your conclusions are correct and present the whole subject of cultivated plants in a new light."[187] In 1855, Harris did write to Gray about Cucurbitaceous seeds (including squash and pumpkins) that he had ordered from Paris, and noted his disappointment that he had not received "some of the kinds that are important in the study, comparison, and identification of species." By this date, he felt he could ask Gray whether he had any botanical contacts in Paris and if he could get the seeds for him. His communication to Gray also included critical remarks regarding European work on the plant group in question as well as general comments on introduced species (with examples).[188] Whether or how Gray responded to this request has not been determined.

INVOLVEMENT IN SCIENTIFIC ORGANIZATIONS

The conduct of science required not only personal relationships, but involvement also in formally organized bodies for mutual support and promotion of scientific study. Harris's memberships reflected the nature of his career and his scientific and scholarly interests. He was a member of the Massachusetts Medical Society (1823),[189] the American Academy of Arts and Sciences (1827), and the Massachusetts Society for Promoting Agriculture.[190] His first publication (on the saltmarsh caterpillar) in 1823, as discussed in an earlier chapter, won a prize from the agricultural society and appeared in its journal. Harris was a corresponding member of the Academy of Natural Sciences of Philadelphia (1826), and his election was an event that he took quite seriously. At the time, the Academy of Natural Sciences suffered from a lack of materials for publication and Harris attempted to assist by submitting a paper on the scarab beetle genus *Cremastocheilus*. He encouraged Hentz to contribute to the academy's journal as well, either on his own or by assisting Harris, writing, "In making this proposition I am actuated by an esprit du corps, a desire to uphold an institution which is in want of support."[191]

Prior to the establishment of the Boston Society of Natural History, Harris's scientific interests were directed especially to the Massachusetts Horticultural Society. Along with his father, he was an original member of the Horticultural Society when it was founded in 1829 and that year he was chosen one of the thirty-eight councillors (or trustees), serving until 1835. In 1829, he also was chosen as a member of the Committee on the Library and from 1829 until his death he had the unpaid position of professor of entomology. (At the same time, Jacob Bigelow was appointed professor of botany and John White Webster professor of horticultural chemistry.) In its founding year, the Horticultural Society voted to arrange with Harris to prepare an exhibit on insects injurious to fruit trees for permanent display in the society's hall.[192] Harris's anniversary address to the society in 1832, on the relations of insects and plants, still considered as a notable event in the organization's history some fifty years later, has been mentioned.

Sometime before 1838, as noted earlier, Harris delivered lectures to the Dorchester Lyceum on the physiology of plants and "analytical botany,"[193] and in the early 1830s he was particularly active in lecturing on entomology before the "Dorchester and Milton Lyceum" and elsewhere in Dorchester and Roxbury, and to the Cambridge Lyceum.[194] In 1840, he received another invitation from the Dorchester Lyceum, but it is not known whether he accepted. If so, the lecture would have earned him fifteen dollars.[195] In August 1842, the Boston-based Society for the Diffusion of Useful Knowledge asked him to deliver three or four lectures "on the scientific subjects which are most interesting to you."[196] Harris gave two zoological lectures for the society in February and March 1843, on the "structure and habits of various animals," for which he was paid a total of eighty dollars.[197]

The Harvard Natural History Society was founded on May 4, 1837. It is not known whether Harris had any role in establishing this student organization, but he became one of its important supporters. In his history of the society, Edward Everett Hale (A.B. 1839) noted that, when he was a member, the chief interest was in ornithology and entomology.[198] During the society's early years, Harris was the college instructor in natural history and had an optional Friday evening lecture on entomology.[199] His nineteenth-century biographer, Thomas Wentworth Higginson (A.B. 1841), remembered that students had access to Harris's insects when they were working on the society's collection.[200] Hale likewise appreciated the special role that Harris played in its early history, reminiscing, "All of us were encouraged and helped by the advice and instruction of Thaddeus William Harris . . . He would be just as enthusiastic about our discoveries as if he had never seen the like before; he would direct us to localities, name our specimens for us, lend us books, and refer us to anything that was in the library with a kindness and patience beyond praise." On May 4, 1838, Harris gave the Harvard society's first anniversary address, of which Hale remarked, "I well remember that we were all so much indebted to him that we wanted to mark our obligation."[201] Harris's assistance to the society continued and he wrote in 1845 to President Josiah Quincy in support of a request by the students for additional space: "Believing the object of the Society to be good, and the collection worthy of preservation & likely to become more & more interesting & valuable as College property,—I hope that you will be able to comply with the request of the petitioners."[202] Harris was also a member of the Cambridge Scientific Club which was founded in 1842 and consisted largely of university faculty members.[203]

Harris's place in the national entomological community was recognized in 1842 when he was elected the first corresponding member of the Entomological Society of Pennsylvania.[204] John G. Morris was one of the chief founders of the Pennsylvania society and had written to Harris several years earlier indicating the salutary effect that a letter from Harris could play in encouraging entomological study among isolated students of the subject. The two men had also shared positive sentiments about the benefits of a meeting of American entomologists.[205] Among the founders of the Pennsylvania society was Harris's correspondent Frederick Ernst Melsheimer, who was made president.[206] Publication of a catalogue of American Coleoptera was one of the chief projects for the new society, which was discussed above, and Harris was asked to contribute "short specific characters of the uncharacterized species in your printed catalogue."[207] Harris's response to the notice of his election indicates the pleasure it had given him. "Your plan is very good; &, in imitation of your example, I shall try to induce some of the young men here, who have become interested in the study of insects, to form an Entomological club (I like this name better than Society), in the hope that, like yours, it may tend to keep up their interest & help create & extend a taste for such pursuits." He further noted,

"Should it be in my power seasonably to send to you, or to Dr. Morris, the characters of the supposed new species of Coleopterous insects named in my Catalogue, I shall be happy to do so, & will forward also specimens for examination; but I am sorely pressed for time, being very much occupied by duties in the library, & having to prepare two lectures to be delivered in Boston this winter." He also included a table of certain insects and promised others, which he had drawn up and which he thought might "serve to amuse if it does not help you." At least one of the tables was noted as having been done some years before, so he was, in effect, drawing on his long-term efforts in order to be helpful to the Pennsylvania entomologists.[208] The initial formation of the Entomological Society coincided with the period of great trial in Harris's professional and personal life. Subsequently, an eye inflammation made it impossible to take part, in regard to which Melsheimer wrote in late 1844, "Our Catalogue of the Coleoptera of the U.S. is progressing slowly to its conclusion, and in consequence of its compilers being deprived of your powerful aid, it will in many instances be deficient in correctness."[209] Later, Harris called on the Pennsylvania society for help in providing information on the transformation of certain insects while preparing the revised edition of his *Treatise*,[210] but there is no evidence that any response was made. In 1845, Harris also was elected a corresponding member of the Entomological Society of London.[211]

INVOLVEMENT IN SCIENTIFIC ORGANIZATIONS: BOSTON SOCIETY OF NATURAL HISTORY

After1830, the year of its founding, the Boston Society of Natural History was likely Harris's primary institutionally based scientific activity. Aspects of the relationship have been given earlier, including Harris's successful effort to procure the entomological collection of Nicholas Marcellus Hentz for the society, his need for a publication outlet as an incentive for establishment of the society's *Journal*, and the society's receipt of Harris's insect collection and library after his death.

The Boston society was formally organized in late April and early May 1830.[212] Harris was nominated on May 13 as a corresponding member but later as a regular member and formally elected to membership on September 2, 1830.[213] He served from May 20, 1835, to May 2, 1838, as a curator, during a time when no departmental assignments were designated; from the latter date until May 3, 1848, he was curator of insects.[214] In the semicentennial history of the society, Harris is noted as among the most frequent contributors of communications during the years 1836–40, and in the first decade he was one of the members most active on society committees. Paralleling the general decline of his naturalist career, he did not sustain the status of frequent contributor during the decade of the 1840s, but he did serve on standing committees of the society's council through which a good part of the society's business was carried out. His active participation in society affairs apparently ceased during the 1850s.[215] Having suggested the value of a society journal in 1832, he was appointed to the publications committee in October 1833 after the original committee was discharged due to the inability of its members to attend to the business.[216] In April 1837 he proposed and served on a committee to address the question of publication of the society's proceedings; the committee recommended such publication.[217] By virtue of his office as a curator, he also served on the society's council.[218]

Harris took his duties as a curator seriously and appears to have devoted a good deal of time to the work. Some of his contacts with other naturalists for exchange of insects were done on behalf of the society, and there were occasions when his transactions were a seamless combination of his own and society interests. For example, in 1838 he wrote that the society's lack of duplicate insects necessitated inclusion of some from his own collection in an exchange with the Swedish naturalist Dom Olof Immanuel Fahraeus.[219] In fulfilling his curatorial duties, it appears that society insects were sent to Cambridge so that he could work on them there.[220] It was a time consuming task, and he noted to Augustus A. Gould in 1837 that "Not a week has passed since I began to work upon the collection of the Society without my doing something about it. . . . you must recollect that I have to determine many of the species & refer them to the new genera—& I can only devote a part of Fridays & Saturdays to this object. In the evening I can do little or nothing about the insects."[221] In addition to time, he made other sacrifices to the effort. In 1836, he wrote that a supply of South American softwood he had procured for use in his insect boxes had to be used for the cabinet of the society instead. Consequently, he ordered an additional ten dollars worth for both himself and the society.[222] In spite of his efforts to get help in developing the society's collections, none was forthcoming.[223] He explained to Edward Doubleday in 1840, "I cannot induce any one of the Society to take up Entomology," and urged the English entomologist to come and join the Boston community of naturalists.[224]

One of the first activities of the Boston Society of Natural History was to arrange lectures on various topics. In September 1830, it was reported that of the various persons asked to lecture, all had agreed except Harris, who had not responded by the time the committee reported.[225] He did, in fact, deliver at least two entomological lectures in the lecture room of the Boston Athenaeum, the first on January 25, 1831. The initial address took a broad approach to the anatomy, structure, functions, and transformations of insects. Harris pointed out that he differed from many other naturalists, dismissing the idea of higher and lower or perfect and imperfect animals as not representing a "truly philosophical conception of natural objects." The lecture apparently was quite well-received and was the basis for other popular lecturing he did around Boston in the winter of 1831 and perhaps later.[226] Later that year he was asked to give three lectures on insects, the last of a series of fifteen, but he wrote that he would be unable to carry out the assignment,[227] probably because of his appointment as Harvard University librarian.

In the winter of 1833–34, he gave three lectures (for which he was paid the society's rate of fifteen dollars per lecture)[228] on the natural history of insects, and especially on their transformations, which were delivered in the Masonic Temple. The final lecture, and apparently the last in the society's series for the year, included a biographical-historical review of the study of American zoology and a solicitation of donations to the society's collections.[229] He reported somewhat later that "The lectures which I was called upon to deliver before the Natural History Society in Boston gave a different direction to my studies for a while,"[230] but he does not elaborate.

In 1840, a particular distinction came to Harris when he was asked to give the annual address to the Boston Society of Natural History. The invitation from David Humphreys Storer, on behalf of the committee, noted that they "feel that no member is better known as a scientific naturalist," but Storer did not mention that the occasion was the society's tenth anniversary. The manuscript of Harris's address ranges over the whole field of natural history; among his assertions was a belief that "species alone have been created. Classes, orders & genera, are mere contrivances of man's invention," and his view that most American species are distinct from those of Europe. In regard to the local scene, he commented favorably on the growth and arrangement of the society's collections and the number of visitors who came to view them. Concerning the state survey then in progress (of which he was a member), he noted that its publications had not "sacrificed science to mere popularity," while reminding the society of its obligation to develop an interest in natural history among the general population.[231]

5

Prejudices and Principles: Ideas That Guided Living

POLITICS AND SOCIETY

HARRIS'S AVAILABLE CORRESPONDENCE IS LARGELY ENTOMOLOGICAL, and it is clear that it was the central focus of his professional life and interests. If he was not actively involved in public events, he nonetheless had strong views on political and social matters that reflected his Federalist origins and his adult years as an ally of the Whig Party. Underlying his political views was a deep concern for the agricultural future of New England, and a personal distaste for the growing immigrant population that revealed a darker side of an otherwise gentle and judicious man. His awareness of public events as well as his sense of humor were displayed in a letter to Nicholas Hentz in 1827, at a time when the governor of Georgia sent state militia in opposition to federal troops that had been detailed to prevent the surveying of Indian lands.[1] Hentz at the time was living in North Carolina, and Harris wrote that Charles Pickering "informs me that he has sent you a 'detachment of spiders,' but has not heard 'whether they have joined you.' I hope you will not let them escape into Georgia, least they should be mistaken by the rebels for some of Uncle Sam's troops sent to suppress the insurrection, or be seized & executed as Spi-es."[2]

Harris's political views, as with his sentiments regarding his entomological career and his overall life situation, were shared in writing especially with Edward Doubleday in England. In April 1840, he wrote to Doubleday, "In your letter you allude to the sufferings of the lower classes in England. Why do not these poor people come to the United States. Hitherto we have had (with but few exceptions) only the very scum & refuse of Ireland & of the English paupers. These we do not want, for they become a burden upon us, & have too many bad propensities & vicious habits. But we do very much want the virtuous, industrious, & active peasantry of England. Young

females, especially, <u>will better their condition in this country imme-
diately</u>." He explained that low wages prevailed at the time and
therefore would increase in the future, benefitting the immigrant
worker. "We still however have to pay enough for <u>female help</u>—our
maid of all work, a young girl from Maine, receives one dollar & a
half per week, & has a very easy time you may be assured, as we enter-
tain no company, & do much of our own work ourselves. . . . She will
be married before long, & then we shall be in trouble to supply her
place, for good girls are very scarce, and I will not have an Irish or a
Catholic girl, if it can be avoided."[3]

Shortly thereafter, Harris wrote again to Doubleday with a fuller
display of his political outlook.

> Our country seems to be undergoing a political change. Present appear-
> ances indicate that the dynasty of Jackson & Van Buren are drawing to
> an end. I wish the views entertained by the Whig Party were more pure &
> consistent. . . . I am sorry . . . to find that some of the best of our northern
> Whigs are now advocating measures entirely opposed to the views of the
> old federal party from which it was to have been expected they would
> never have greatly departed. . . . By advocating high wages as they now
> do, they seem to be appealing to the worst passions of those persons,
> who, having no real property, have no interest in maintaining good
> order in society. High wages have been greatly injurious to our farmers,
> the main stay of the country, & have tended, heretofore very much to
> depress Agriculture among us. High wages & high duties have tempted
> us to go largely into the manufacturing business to the neglect of agricul-
> ture. . . . When the native productions of our country can be raised
> in such abundance as to supply all our wants & leave us a surplus for
> exportation, then may wages be raised without detriment to the farming
> interest & then too will it be wise for us to turn our attention to manufac-
> tures.[4]

Here Harris displays a political foundation for his interest in agricultural entomology, a view that he did not often have occasion to express. Something of the same sentiment, however, was given publicly
in the preface he wrote, on October 15, 1852, for the second edition
of his *Treatise*, as a justification for the form that that work took. "Believing that the aid of science tends greatly to improve the condition
of any people engaged in agriculture and horticulture, and that
these pursuits form the basis of our prosperity, and are the safe-
guards of our liberty and independence, I have felt it to be my duty,
in treating the subject assigned to me, to endeavor to make it useful
and acceptable to those persons whose honorable employment is
the cultivation of the soil."[5] Undoubtedly his extreme prejudice

against Catholics, and the Irish in particular, was derived from multiple factors, but even to that view he gave an agricultural emphasis, noting that "they will not disperse through the country, nor employ themselves in agricultural pursuits."[6]

As suggested above, Harris was a Whig, but he was not in lockstep with that political body, admitting "though I have never voted against the party, I have sometimes declined going to the polls." While his politics seem historically traditional in his time and place, Harris thought his opinions might "be visionary & impracticable at present," but pointed out that they were the result of careful thought over time.[7]

Harris opposed the war with Mexico[8] and expressed himself, at least indirectly, on the question of slavery, on one occasion stating that the so-called Cotton-whigs[9] "will sacrifice the cause of emancipation for the slave and the perpetuation of freedom, to the so-called protective policy."[10] His sentiments on the subject also were revealed in another context. In 1848, S. W. Cole, at the *Cultivator* office in Boston, wrote to Harris and included insects that had been found on potato plants. On the back of the letter is an unsigned draft of a manuscript inspired by the shipment, and other drafts are in Harris's papers. The most finished form of this piece, undated as are the others, but with the note "For the Boston Cultivator," reads

> "Lady-bird! Lady-bird! fly away home!
> Your house is on fire, and your children will roam."
>   Nursery Rhyme

A whole family of these little children have come to me; not voluntary wanderers, or fugitives from a burning house, but as captives, taken in their native haunts, shut up in a dark box, like Africans in a slave-ship, and sent whither they know not, at the risk of starvation and death in their passage, and threatened with a fate worse than death itself. But the poor captives have fallen into the hands of one who knows how to pity & relieve them, and who, hating oppression and slavery in all its forms, and hoping that soon every land may be declared forever "a free land," & the whole world "a free world," will hasten to give them their liberty.

You would know of me something of their history; and I will tell you, as you wish it. of their parentage, and how they lived, what changes they have undergone in their free state, and what still await [*sic*] them.

The lady-birds pass the winter concealed in some crevice, and come forth in the spring to prepare for a future family. During the summer they are abundant in gardens, fields, & orchards, attending upon plant-lice, which are of the greatest importance to them. There are several kinds of them, but seldom more than one kind on any one plant. Those that I have liberated inhabited the potato. To the lady-bird and her family a potato-field is an endless forest, and a single potato-vine "the bush" where they make their home. The parents bear a striking resemblance to each other, and a description of one will serve for either of them. In form & size they are like the half of a split pea, being convex on the back & flat beneath. The shelly covering of their back is of a dull orange-red color, specked with a few black dots; and the body beneath is a greenish black. Their legs, of which they have six, are also black. They creep slowly, but fly well, having a pair of long filmy wings, folding under their convex shells. When touched they give out an unpleasant smell, that comes from juices oozing from their joints. The female nestles beneath a leaf, and lays there a cluster of several golden yellow eggs, which are generally hatched in the course of twelve or fourteen days.[11]

This interesting document combines a lightness of tone with humane convictions, and imparts entomological information written in such a way as to be maximally accessible to the general reader. Although uncharacteristic of his writing, it suggests how much his fundamental political sentiments and scientific interests were integrated into his outlook on life.

Religious Practice and Belief

There is surprisingly little direct reference to religion in Harris's correspondence and other papers, and the biographical accounts by his son and daughter are equally silent on the subject.[12] The fact that Harris's father was a Unitarian clergyman makes the absence of religious sentiment particularly puzzling. His father's view of religion was seemingly ecumenical, and he was said to have been more concerned with the practical than the theoretical in his religious orientation,[13] which might help to explain Harris's apparent reluctance to commit to strong views in religious matters. Only fragments of evidence have been discovered about his religious thought and practice. In one of his earliest extant letters, written while visiting New York, he reported to this sister that he had gone on Sunday to the Congregational Church, "or Unitarian Chapel as it is here called."[14] From this it appears that he was in the habit of attending church, since he did so while traveling. Some years later, in writing to his correspondent friend Thomas Affleck in Mississippi, he combined humor with a suggestion that church attendance competed with other uses of his precious free time.

This is Thanksgiving-day in the land of johnny-cakes & pumpkin-pies. We had a North-east storm yesterday evening, & the ground is covered this morning with two inches of snow. My boots, alas, are too hole-y to go to the house of worship, & as I cannot follow the recommendation of our Governor without them, I must stay at home,—and am glad to have & to improve the holiday in acknowledging receipt of your insects, & to make some remarks thereon.[15]

By all evidence, Harris continued his denominational affiliation as a Unitarian throughout his life. He was listed as a member of the First Parish, Unitarian, in Cambridge, Massachusetts, on January 1, 1854, in a special category labeled "Admitted on Notice Name not on assessors list."[16] When his young son William died in October 1854 at age twenty-eight, the funeral was conducted in the First Church (First Parish) in Cambridge,[17] and Harris's own funeral, on Sunday January 20, 1856, was officiated over by the minister of the same congregation.[18]

Beyond the matter of organizational affiliation, Harris's specific beliefs are divulged only in incidental remarks that are expressed in unoriginal ways and made chiefly in relation to life events. Thus, in consoling Nicholas Hentz in 1827 on the death of a young son, Harris wrote, "if the chastisement of Providence is in your case uncommonly heavy, may [you?] be supported & comforted by the assurance that 'of such' as him 'is the kingdom of heaven'."[19] Some years later, he wrote of two elderly relatives whose deaths he mourned and was consoled by the expectation that they had gone to their "heavenly inheritance."[20] Both comments indicate his sense of Providence and the reality of a life beyond death. Although a Unitarian, his obituaries show a denomination still tied to its historic roots,[21] so that one writer noted, "He . . . has died as the Christian dieth, leaving behind him the savor of a good name."[22] A notice written at the time of his death, probably by his Harvard classmate John Gorham Palfrey, said of Harris, though "Of unrivalled eminence in his walk of science, he was too diffident of himself to admit of others doing him full justice. Yet he had abundantly the self-respect which belongs to unselfish labors to advance the world in the knowledge of the works of its Maker, and to the uniform tenor of a pure, useful, Christian life."[23] In fact it is in passing references to theological or spiritual themes in his entomological writing that real glimpses of Harris's religious views are found.

RELATIONS TO NATURE AND SCIENCE

Harris's occasional references to religious values in his *Treatise* are expressed in conventional terms of providence and the wisdom and benevolence of God through design in nature—derivative from the natural theology that was so significant a part of religious sentiment in that time period—yet there is no reason to think they were presented with anything other than genuine attachment. He wrote, for example, of the relation of the Baltimore oriole and the pea-weevil: "The instinct that enables this beautiful bird to detect the lurking grub, concealed, as the latter is, within the pod and the hull of the pea, is worthy our highest admiration; and the goodness of Providence, which has endowed it with this faculty, is still further shown in the economy of the insects also, which, through His prospective care, are not only limited in the season of their depredations, but are instinctively taught to spare the germs of the pease [sic], thereby securing a succession of crops for our benefit and that of their own progeny."[24] The use of theological argument in nineteenth-century science has been referred to both as part of the justificatory stance of scientists in seeking social accommodation of their interests, and as reflective of the close relations between science and religion in the first half of the nineteenth century.[25] Robert Bruce has concluded that "the religious asides of most antebellum scientists were manifestly sincere,"[26] and this most certainly applied to Harris.

Harris commented further on the relations of animals, humans, and the divine plan when he wrote, following discussion of insects parasitic on other insects: "Such are some of the natural means, provided by a benevolent Providence, to check the ravages of the destructive Hessian fly. If we are humiliated by the reflection, that the Author of the universe should have made even small and feeble insects the instruments of His power, and that He should occasionally permit them to become the scourges of our race, ought we not to admire His wisdom in the formation of the still more humble agents that are appointed to arrest the work of destruction."[27] He also pointed to physical features of certain insects that worked for their defense against enemies by rendering them invisible in their environment. In noting such providential care taken for the preservation of these "most despised of God's creation," Harris took the opportunity to argue divine concern as a reason for humans to study insects. This is only a momentary interruption in his text, however, for he quickly announces, "But to return to our locust tree-hopper, which remains to be described . . ."[28]

There is a general lack of aesthetic expression or religious sentiment in Harris's personal and professional letters, which suggests that it did not come easily to him. The interspersing of occasional comments on creation and design in nature in his published work, therefore, probably was intended to have a certain rhetorical role in promoting the study of nature and, perhaps, to underscore in one more way his fundamental premise (central to his practical entomology) that understanding the complexities in the workings of nature was essential to controlling it.[29] Irrespective of an ulterior intent, however, there is every reason to believe that the religious sentiment in the *Treatise* is given with sincerity, as well as with conviction. In any event, it is not often introduced.

Poetry is not expected from one of Harris's sensibilities, but on at least one occasion he did indulge an interest, resulting in a manuscript that is particularly telling about the function that nature played in his personal or psychic life. It is beside the point, for biography, that it was not particularly good poetry. At the age of twenty-seven, two years into his medical career and not yet married, he composed "Lines written on the author's Birth-day Nov. 12, 1822." It is a seasonal piece that notes the passage of the flowers of spring and the summer sun, and even "Sober Autumn almost gone." Yet Nature's charms have still the pow'r / To banish sorrow for an hour," "Its varied wonders yield relief / To one o'erwhelm'd with care & grief." In this melancholy state, either in his mind or in fact, he went to the woods and encountered "The mystick Hazel's slender spray," whose yellow blossoms were a saving presence in the emptiness of the autumn trees. The manuscript ends (almost) with a sense of gratitude for such an encounter with nature but also with disappointment that the sentiment is not universally felt, for, "Though Nature smiles on all around, / Who feel her smiles how few are found. / And though her bounties wide extend, / How few to own her for a friend." Even as a young man, however, Harris's temperament was such that he could not leave poetry to carry its own weight, so that after the poem he adds a footnote on "The mystick Hazel."—"The 'Witch-Hazel' puts forth a profusion of yellow blossoms in November, & forms its fruit the Spring following. The twigs were formerly used, for the purposes of imposture, in diving [divining] for minerals."[30]

From whatever motivation, for whatever purpose, Harris seems to have been in love with nature and with nature study. As indicated in earlier sections, his opportunities for direct study and extensive contacts with nature varied and diminished, overall, through the course of his life. After his move to Cambridge in 1831, his time and opportunity for direct entomological study lessened, and he became dependent on others for additions to his collection and for specimens useful for exchange.[31] While working on his State Report in 1840, he wrote to Frederick Melsheimer that for eight or ten years he had not had much opportunity to engage in direct study of the history and habits of insects, so that "almost all that is known to me from personal observation was learnt a long time ago."[32] His situation did not improve thereafter, telling John Lawrence LeConte in 1846 that a particular insect was "not common in Milton, my former place of residence, where most of my insect collections & observations were made."[33] If the range of his opportunities contracted, he was prepared, nonetheless, to take advantage when a situation presented itself, as in 1836 when he found many caterpillars on his grape vines, from which he "obtained several fine moths from larvae which transformed in pots."[34] His opportunities for collecting also took on a decidedly domestic aspect, noting to Nicholas Hentz that "During warm summer evenings I catch many small Lepidoptera on the walls of the parlor, by putting a wine-glass over them."[35] Field collecting was not entirely out of the question when he had a special need, so that in the summer of 1838 he spent a morning on the marshes of Cambridgeport in a futile search for a particular insect, and a few days later went to the Blue Hill for another and met with greater success.[36] In 1839, he wrote to Doubleday that "The autumn has been remarkably fine & mild, & I have improved it in making short excursions every Friday afternoon & Saturday with Dr. Zimmerman."[37] In summer 1848, he went to Blue Hill in search of another insect but could find only five specimens.[38]

Harris's particular interest in the life histories of insects may be seen as evidence of both his love of nature study and of science. As opportunity arose, it kept him engaged with living insects rather than devoting his efforts only to his collection. Commenting in 1852 on the relations of various aspects of his entomological program, he wrote, "By far the most interesting & profitable portion of the time given to this pursuit, has been that which has been spent abroad in watching & taking insects in their natural haunts." He then went on to query his young correspondent, "[D]id you ever see a Brachinus clean its antennæ with its fore-legs? It would give you an idea of the use of the deep emarginations in the anterior tibiae, which are admirably calculated to assist in this operation."[39] The publication of his *Entomological Correspondence* in 1869 included a section titled "Descriptions of Larvae: Memoranda of Their Metamorphoses, Habits Etc.,"[40] which includes a number of dated notes on Harris's capture observation, and description of insects over the period 1821 to 1855. Among the insects noted was the butterfly, *Papilio Philenor,* collected in the summer of 1840 and reported to entomological friends as never having been found in Massachusetts before.[41] Of the event Harris's student and biographer, Thomas W. Higginson, wrote "Every great discovery was an occasion for enthusiasm, and i seemed the climax of his life when he found for the first time, on August 5, 1840, the larvae of the southern butterfly, Papilio Philenor, on a shrub in the Botanic Garden."[42] Harris was more subdued when writing about the insect, reporting "the interesting discovery" to Doubleday,[43] and expressing his "satisfaction of finding" the caterpillar to Melsheimer.[44]

In an 1833 letter, Harris admitted "I am never more happy than when investigating some object of nature."[45] A fuller appreciation of his emotional engagement with nature is more likely to come, however, from comments by contemporaries than from his own correspondence or published writings. T. W. Higginson's account of the quest for a certain flower reveals not only aspects of Harris's relations to nature but his approach to natural history study itself.

> I remember the perennial eagerness with which he urged upon us [students], each spring, to rediscover the Corallorhiza verna in a certain field near the Observatory. It had been found there once, and once only, by my classmate, Dr. Woodward. . . . Dr. Harris's eyes would always kindle when the little flower was mentioned, and he would ponder, and debate, and state over and over again the probabilities and improbabilities, and discuss the possibility of some error in the precise location, and draw little plans of that field and the adjoining fields and urge us to the pursuit or cheer us when drooping and defeated, until it seemed as if the quest after the Holy Grail was a thing insignificant and uninspiring compared with the search for that plain little orchid.[46]

Harris's son likewise left his impressions of his father in the field as enthusiast, naturalist, and genealogist.

> Indelibly fixed upon the writer's memory are the recollections of the bright, sunny days spent with him in rambling over the beautiful country about Belmont and Waltham; of the sudden rushes after some flying Buprestis, or the wary chase of some shy Cynthian; of the bark-stripping in search of the Curculio larva, and the search in the meadow pools for the Dytiscus. And sometimes net would be dropped and a stone wall scaled that an old, mossy grave-stone in a forgotten burying-ground could be cleaned and its epitaph transcribed in the note-book. On such occasions every passing insect and every way-side plant furnished a text for such lessons as only he could give. And through it all, ran that thorough delight in the subject, and the tenderness, almost feminine, for everything that the Creator had made, that were so markedly a part of his being.

In his reminiscences, the son noted that the only thing that competed for his father's enthusiasm in entomological study was a love of natural scenery.[47]

Entomologist Samuel H. Scudder (1837–1911), who did not know Harris but edited his *Entomological Correspondence,* wrote that he was "Painstaking and laborious to the last degree in all he undertook, his accuracy has never been questioned and his principal work with its simple, direct style can never be superseded. He was the Gilbert White of New England."[48] While Scudder's comparison of Harris to the author of *The Natural History and Antiquities of Selborne* is an interesting attribution, it is not entirely clear what qualities Scudder intended to convey in doing so. Certainly Harris had none of the broad view of natural history but was, instead, very much a specialist in entomology in spite of his interest in botany. He was oriented to his native region, at least in the initial stages of his natural history interest, and his plan in the 1820s to publish a Faunula Insectorum Bostoniensis evokes the strong sense of place in his scientific work. That focus expanded in the course of his career as his contacts and resources developed, but the editions of his *Treatise* during his lifetime still were limited by title to New England (the posthumous third edition omits all reference to geographical location in the title). Scudder referred to Harris's knowledge of the life histories of insects as the source of his reputation,[49] and that faculty for close and careful observation may be the basis of Scudder's comparison to White. It is a comparison that effectively elevates the literary quality of Harris's published work, but it ignores his equally strong interest in taxonomic studies and the wide range of his contacts and interactions with naturalists in both the United States and Europe. Harris was acquainted with White's work and quotes him in the *Treatise* on the sound of the field cricket. Although White wrote that the cricket's sound "marvelously delights some hearers," bringing to mind rural delights of summer, Harris speaks for the others, countering that these are "sentiments in which few persons, if any, in America will participate; for with us the creaking of crickets does not begin till summer is gone, and the continued and monotonous sounds, which they keep up during the whole night, so long as autumn lasts, are both wearisome and sad. Where crickets abound, they do great injury to vegetation."[50]

Harris's writing was straightforward and committed to the transfer of information. Higginson, in the quotation above on Harris's quest for a once-seen flower, concluded with the comment that "this was the true spirit of the observer,—appreciation of the unspeakable value of a fact."[51] There were occasions, nonetheless, in his correspondence and his published writing when Harris let loose the constraints that were natural to him and to which he would have subscribed, as well, as a participant in the professionalizing scientific community. In one of his letters to Edward Doubleday, written after Doubleday had visited in 1838, Harris expressed his hopes for an ongoing epistolary friendship between the two naturalists, remarked on the status of "right of priority" for taxonomic names in France, and on his desire to promote entomological study in the United States (and Boston in particular). He concludes

Since you have been here we have had an epitome of summer & winter.—For a few days last week the temperature of the air was truly delightful—the Indian summer had arrived—the fields which, as you may recollect, had been uncommonly green during all the autumn seemed fresher & more verdant still, in consequence of previous copious rains— Although the deciduous plants had shed most of their leaves, the bushes & trees were animated by flocks of little birds, twittering & chirping while they gathered their food & plumed their wings for flight. The faded grasshoppers jumped in the paths; & some of the vernal Geometers began to emerge from their winter quarters; swarms of harpy flies crawled out of the crannies of the walls & persisted, in spite of all efforts, to share with us in our repasts; even now and then a solitary Cicada would be heard rattling his discordant drums, & beating the roll for recruits; while often the clangor of the geese would fill the air, as phalanx after phalanx swept over our head—& in the evening the crickets once more crawled from the holes & gave again their parting & mournful serenade. Saturday was the last of these lingering summer days—on Sunday it began to snow—& in the morning a fleecy mantle, two inches in thickness clothed the ground;—the evergreens were so thickly powdered that their branches [hung?] about them like dissheveled [sic] locks & were weighed down almost to the ground—every little twig of the deciduous trees seemed puffed up with vanity; & even the very posts & rails assumed a magnitude & importance which did not belong to them. In fact there was snow enough to set the bells a jingling & all the little boys arriving with their sleds, & to [fill?] all their shoes as they went to school . . . Some of this snow was melted in the course of the day—but most of it remains even now. I have sat up late to write to you—& have just opened the door & looked abroad—all sounds are hushed in the streets—it is perfectly calm—& though it is cool enough to freeze it is not uncomfortably cold even in the open air. . . .—the sky is that beautiful deep cæsious hue which we love in a blue eye, & the stars are looking through the blue vault as though they were coming closer & closer to you—Some of these silent watchmen of the night have gone to rest—& it is time that I should betake myself to my slumbers—This communication with you & with nature has brought my mind to a fitting state for calm & peaceful repose—& that such ever may be yours is the wish of your sincere friend.[52]

This is an uncharacteristic outpouring by Harris, whose letters include few such reflections on the natural world and none as sustained as this. Most certainly there was genuiness in it, but perhaps also an attempt, in other language, to establish rapport and common ground with the English entomologist whose friendship his letter so ardently solicits.

In reference to summer trips to the White Mountains of New Hampshire in the 1850s, Harris's son observed that "His delight in the contemplation of the beautiful and majestic scenery of that locality was ever as fresh and intense as at first."[53] It is doubtful, however, that a sense of the sublime—described as a frequent feature of American reactions to nature during his time[54]—applies in any real sense to Harris, though his daughter, referring to her mother's recollection, reported that he was "the most delightful of travelling companions. He played well;—throwing off all the cares and anxieties of every day life, he would throw himself into all the interests and novelties of travel with the enthusiasm and fun of a boy out of school . . . He had a beauty loving soul and feasted on a beautiful view with deepest satisfaction."[55] Perhaps in a more demonstrative personality more of the emotional encounter with the natural world would have been registered.

In a letter written to Dr. William LeBaron in 1850, Harris combined a travel account with an impressionistic but vivid report of his observations of flowers and insects, giving what undoubtedly was a true rendering of his thoughts and interests, always the naturalist even while a simple traveler.

As this was my first visit to the [White] mountains, it was greatly enjoyed, although in a scientific way unproductive of results of any importance. Throughout the whole excursion [from Maine to New Hampshire] I was struck with the paucity of species of insects observed, with the exception of those of the genus *Bombus,* which were to be seen in great profusion on the Canada thistles. *Antennaria margaritacea* was found in abundance in almost all parts of our route, sometimes covering whole fields with their white flowers, often of immense size and great beauty; but they were without any attractions to the insect tribes, and were rarely touched by them. The fire-weed, *Epilobium spicatum,* threw up a profusion of its showy purple-flowered spikes wherever the wood had been burnt over; but these flowers also harbored very few insects; the golden-rod was just coming into flower in Maine, and in some of the warm valleys in New Hampshire, and where sufficiently expanded had some insect visitors upon it, such as *Lepturae,* etc. A few specimens of the beautiful northern *Buprestis fasciata* were taken on Pleasant Mountain and on Kearsage.

The report continues in this vein and finally, he concluded, "I never before spent so much time in the open air in the summer, with fewer

insect acquisitions. The excursion nevertheless was abundantly enjoyed in other respects; and as the journey from Portland to Franco-
nia was taken in a private carriage, I had opportunity to stop
whenever and wherever I wished."[56]

While Harris clearly was enthralled by natural scenery and by contact with nature on a minor scale, as well as by routine study of natural objects, it would be a mistake to conclude that religious or
aesthetic sentiment was a primary motive for his studies. His motives
were multiple and complex. In spite of the sensibilities toward nature that showed occasionally, and in the face of scruples about pinning live specimens,[57] much of his published output related to the
control of insects, with appended suggestions for their destruction.
A concern with injurious insects meant that nature and humans
were closely linked in Harris's thought. Insects, however, were not
the only destructors. Harris accepted and used the idea of an inherent balance of nature,[58] and in that vein argued,

> too often, by an unwise interference with the plan of Providence, we de-
> feat the very measures contrived for our protection. We not only suffer
> from our own carelessness, but through ignorance fall into many mis-
> takes. Civilization and cultivation, in many cases, have destroyed the bal-
> ance originally existing between plants and insects, and between the
> latter and other animals. Deprived of their natural food by the removal
> of the forest trees and shrubs, and other indigenous plants that once
> covered the soil, insects have now no other resource than the cultivated
> plants that have taken the place of the original vegetation.

The means to correct this destructive situation was knowledge of insect ways, including their relationships "to each other and to other
objects."[59] Specifically in relation to the destructive cankerworm,
susceptible to predators and to being parasitized by other insects,
for example, he observed, "Without doubt such wisely appointed
means as these were once enough to keep within due bounds these
noxious insects [cankerworms]; but since our forests, their natural
food, and our birds, their greatest enemies, have disappeared before
the woodman's axe and the sportsman's gun, we are left to our own
ingenuity, perseverance, and united efforts, to contrive and carry
into effect other means for checking their ravages."[60]

His concern for the diminishment of natural areas was not just
because of the effect on predator–prey relations. In 1850, Harris
wrote to Asa Gray about the disappearance of once flourishing
flowers in Cambridge, where they had formerly been "so accessible,
that, in an early morning's walk, I could always find them there for

my lectures." This loss had consequences that the man who succeeded him as instructor in natural history could appreciate, arguing that the disappearance of "such plants as these [just enumerated], in the vicinity of the Colleges, is to be regretted, for it deprives the young botanical student of one of the chief inducements to the study, namely the pleasure of finding the plants himself." Harris hoped that Gray might be induced to resurrect some of the locally lost plants by cultivating them in the Botanical Garden.[61]

It is clear that the retreat of nature affected Harris strongly, though the surviving references are few. In a letter written shortly before his death to Thomas Wentworth Higginson, Harris lamented the changes that had taken place in the Cambridge area during his lifetime, confessing, "I mourn for the loss of many of the beautiful plants and insects that were once found in this vicinity." He delineated a number of species and larger groups, using their Latin names, plants which "have been rooted out by the so-called hand of improvement," and insects "which for several years occurred in profusion on the sands above Mount Auburn . . . [that] have entirely disappeared from their ancient haunts, driven away, or exterminated perhaps by the changes effected therein." But looking, with hope, beyond Cambridge, he suggested to Higginson, "There may still remain in your vicinity[62] some sequestered spots, congenial to these and other varieties, which may reward the botanist and entomologist who will search for them carefully. Perhaps you may find there the pretty Coccinella-shaped, silver-margined *Omophron,* or the still rarer *Panagaeus fasciatus,* of which I once took two specimens on Wellington Hill, but which I have not seen since."[63]

During his time, Americans generally believed that the earth was intended to be used by humans.[64] In the United States, and especially among the Puritans and their descendants, there was an environmental reform ethic manifest in literary sources from colonial times. But rather than a consciousness of the negative aspects of human action, it was an ethic for improving the land, an attitude derivative from biblical sources.[65] In Harris's work as a practical entomologist and his promotion of agriculture, he saw the use of the land as a positive social good, a bedrock of political and economic stability. The farming population shared that perspective; they were by necessity interested in how nature could be made useful and productive. But at the same time farmers had a stake in preservation along with improvement, and their relations to the land and to nature had a spiritual and moralistic facet along with the practical. Nature inevitably had injurious and beneficial elements, and natural history of a populist strain was important in the agricultural sector.[66]

To maintain the inherent relations in nature, and farming in the political economy of the nation, Harris argued for the restoration of a natural system that had been disrupted by human agents. He, and others who argued from the perspective of an original balance in nature, were aware of the damage done by past action. In his considerable effort to develop and disperse information on methods for eliminating insects that were injurious to agricultural interests Harris, therefore, had to hold in view both the negative effects of human action in disrupting a natural balance and the need to take positive steps to protect plantings when the inherent checks did not suffice. As such, he participated in the ambivalent relation to the land that was inherent in the pursuit of agricultural interests.

Scudder is no doubt correct that the reservoir of Harris's relations to nature was his deep interest in the study of life histories of insects and the relations of flora and fauna that such an orientation inevitably entailed. In the middle course of his scientific work—especially in the 1830s and early 1840s, when he became enveloped for a time in the taxonomic effort that motivated so many practicing naturalists, the years when he tried to establish a place for himself in the international community through his studies of the nocturnal Lepidoptera—one might expect a dilution of that sense of nature, that the life of the insect would be sacrificed to the efforts of the closet naturalist. But, as discussed in an earlier chapter, he insisted on making life histories an integral part of his taxonomic work and thus maintained that connection to events in nature as well as to physical form. The place of life histories in nineteenth century natural history has not been systematically explored, and there are some contradictions in the result. In England, it is said, there was little or no interest in the study of the habits and behavior of living animals. William Kirby and William Spence devoted the first two of the four volumes of their important *Introduction to Entomology* to life histories, presented in a particularly engaging manner, but ostensibly for the ulterior purpose of enticing readers to the more serious technical sections that followed.[67] Other historians have argued that life history studies were "central to natural history" (although the degree or manner in which this was true may have reflected national differences in emphasis or practice). Thus, in Germany, life histories in the 1830s were considered one of the two branches of natural history, taxonomy being the other. In subsequent developments beginning within Harris's lifetime, life histories were brought into the newer and ostensibly more scientific study of zoology, while description and classification (as well as applied studies) were left outside.[68]

Margaret Welch's study of nineteenth-century America points to life history studies of animals—"virtual biographies" that encompassed "appearance, reproduction, communication, habitat, life stages, and geographical distribution"—as a characterizing feature of natural history in the United States during that period. American naturalists used the approach as a way to exploit their advantages, in the face of a lack of collections and library resources compared to those available to Europeans.[69] Harris's early interest in the study of life histories undoubtedly came, in part, from the lack of information sources on insect description and names, although it appears also to have been genuine and important in itself and an integral part of his programs for insect control and classification. Harris shared his fellow naturalists' idea, that what was a negative result of geography could be turned to a positive value, that is, that life history studies were among the advantages American entomologists had over Europeans. As he argued the point, "We have the country before us; we have the best of opportunities for studying the habits of insects, for detecting varieties, & for discriminating true species."[70] Domestically, he thought it was a way for others to get personal pleasure and contribute profitably to public knowledge. In writing to a young aspirant, whose use of entomological nomenclature Harris had criticized, he pointed out an alternative route: "The field for observation is an extensive one; and those who are disposed to study the habits of insects, will meet with ample reward, as well as be doing great service to the cause of a department of Entomology which has been greatly neglected since the time of Réaumur & De-Geer."[71]

Welch's book gives special place to Alexander Wilson's *American Ornithology* as progenitor of the life history-mode in early nineteenth-century American natural history.[72] Life history as characterized by Welch, "with its usual accompaniment of personal interjection and interpretation," came increasingly to be divorced from technical studies and was relegated to popular writing, at least in ornithology.[73] But the situation with Harris and his approach to entomology was more complex—his interests began with life histories, he took a broad approach to taxonomy so that life stages and insect relations to plants were a significant consideration, and the sources of support for his entomological publishing eventuated in a popular work that emphasized practical values. In effect, with Harris, the several aspects of his entomological effort were seamless and derived from a genuine love of nature and nature study.

Overall, Harris's entomological output was a tapestry of personal study and observation, reports and specimens from correspondents, and published sources. Though never as extensive as he hoped, Harris was effectively connected to the literature of entomology and agriculture and in his later years, as discussed earlier, he devoted some efforts to historical studies of domestic plants. His *Treatise* is well supplied with actual and implied references, summaries, and observations from the published literature, including monographs, encyclopedias, journal articles (general, scientific, and especially agricultural), and newspaper accounts. Although Harris derived obvious pleasure and fulfillment from his physical engagement with nature, in reality and by his own admission he was a closet naturalist, especially after he went to Cambridge and no longer had the time to collect insects very extensively in the field. But he never accepted that as the appropriate role of the naturalist, and as he told his friend Charles Zimmerman in 1837, it was his hope to untether himself from his other responsibilities so that he could go forth and collect more freely and extensively.[74] This did not happen, of course, and his brief vacation excursions into the White Mountains in the 1850s, and occasional outings around Cambridge, perhaps came closest to that fond expectation.

To delve into Harris's being in order to uncover an ideology formative for his life is, in many respects, a distortion of that life. Strands of belief are found and have been laid out above. His political and social allegiance to agricultural interests have been suggested through his own words, and the likelihood that they underpinned certain aspects of his entomological work are indicated. Some aspects of Harris's political views sounded more Jeffersonian than Whig—for example, giving priority to agriculture and support to free trade. But they make more sense when it is realized that participants in the industrialization of New England also felt ambivalent about the change they were bringing, and representatives of the older Federalist faction (based on mercantile interests)—with whom Harris identified—continued to oppose the tariff and to give high value to agriculture. John Lowell was among them, and Josiah Quincy, who as Harvard president hired Harris as librarian, would not invest in industry nor endorse the protective tariff and carried out extensive experiments on his Braintree farm.[75] Thus, and not surprisingly, Harris's ideological perspective has to be seen both in relation to other contemporaries and how he adapted it for his own life-orienting purposes.

Overall, Harris was a thoughtful man, conservative in his views and his demeanor (including his scientific work).[76] His promotion of a restoration of balance in nature has a counterpart in his wish for the maintenance of order in society, and these give coherence to his character. But he was a man devoted to work and, as the quotation from the memoir by his son in the next chapter indicates, his accomplishment was achieved by constant labor.[77] It was on detail, on fact, and the relation of one fact to another, that Harris's ideology, if it can be so expressed, was hinged. Almost at random, a selection from his entomological correspondence shows how his thinking went in practice. To John Lawrence LeConte, 1851:

> You ask me what differences I find between Cicindela rugifrons and C. unicolor. The insect that I refer to the unicolor of Dejean, Vol. I, pp.52, 53, appears to me to be distinct from . . . rugifrons. Numerous specimens and varieties of the latter I took on the sands beyond Mt. Auburn in Watertown, in August, 1826, and subsequently there and at Chelsea Beach, and have seen others from Martha's Vineyard. A practiced eye would at once separate the southern unicolor from all those varieties. . . . It is easier to perceive than to describe the difference between these varieties of rugifrons and the southern unicolor. The latter was taken at St. John's Bluff, East Florida, in February, 1838, by my lamented friend Mr. Edward Doubleday, who kindly presented two specimens to me. The unicolor referred to is absolutely immaculate, not a vestige of white being upon the elytra. The color is more blue than green, with a beautiful purple or reddish reflection on the elytra. Size smaller than rugifrons; thorax more than smooth. . . . All these, I admit, are on paper slight distinctions; but were you to see the specimens side by side with rugifrons, I think the differences would strike you as they did both Doubleday and myself.[78]

And thus it goes—there is reference to bibliographic authority, data on the time and place of his own collection and observation, appeal to the role of experience but admission that subtleties are involved in reaching a conclusion, precision but vivacity as well in describing the specimen itself—all in the context of a finely focused attention and in the interest of promoting his science and (in this instance) furthering the knowledge of an entomologist thirty years his junior. Harris's nineteenth-century biographer, Thomas Wentworth Higginson, was quoted above in regard to Harris's devotion to facts.[79] It *was* the fact that riveted Harris's life and gave it form and content. The effort itself undoubtedly brought its share of fulfillment and for Harris personally it could be one worthy end, but facts also were marshaled toward achieving greater understanding, and to solving both technical (i.e., agricultural) and societal problems. In viewing the larger concerns and in making the argument, appeals were made to religious values (as the role of a creator God who had provided the wonders and the system of intricate design and sustaining interdependence that careful study had uncovered), and to the usefulness of knowledge in taking command of one's destiny. In various places, and through his own practice, he also promoted the values of national self-interest through the development of American science. Through much of Harris's entomological career, there is the tension between science as study and science as knowledge for a purpose. It was a relationship that he had to contend with throughout his life, as ideology and practice rose and fell as salient factors in that life's course and output.

6

Death and Assessments

Death

Harris's last day in the library was on November 8, 1855, four days before his sixtieth birthday. The following day he reported to Assistant Librarian John L. Sibley that he was unwell, and he never left his house again.[1] On November 20, Harris wrote that he had gotten up from a sick bed to answer a letter, commenting, "I have been suffering above a week with acute pleurisy,—and though the cutting pains are removed, there remains much oppression, & shortness of breath, and an effusion of serum in the cavity of the chest."[2] On the morning of January 16, 1856, Sibley visited him for about fifteen or twenty minutes regarding library business, and later reported that "His mind was as clear, his plans and purposes as good, his inquiries, suggestions, and directions as minute; and his confidence that he should soon resume his duties as strong as at any time during his sickness." Probably before Sibley reached the library after that conversation, Harris was dead. His passing was attributed to pleurisy succeeded by phlebitis.[3]

Self Assessment

Harris was not often given to deep self-analysis, at least not in the surviving documents.[4] His dissatisfactions with the course of his career were noted in previous sections, including his aspirations for employment that would support his entomological activities more effectively than medicine or librarianship.[5] In 1839, he confessed, as well, to a sense of isolation and uncertainty. To Doubleday he wrote,

> You have never, and can never know what it is to be alone in your pursuits, to want the sympathy and the aid and counsel of kindred spirits; you are not compelled to pursue science as it were by stealth, and to feel all the time, while so employed, that you were exposing yourself, if

discovered, to the ridicule, perhaps, or at least to the contempt, of those who cannot perceive in such pursuits any practical and useful results. But such has been my lot,—and you can therefore form some idea how grateful to my feelings must be the privilege of an interchange of views and communications with the more favored votaries of science in another land.[6]

The image of the entomologist during this time was ambivalent. In Great Britain, William Kirby and William Spence noted in their *Introduction to Entomology* (1815) that insect study was viewed as lacking in seriousness of purpose, while another English entomologist, Edward Newman, remarked somewhat later that "the collector of insects . . . must make up his mind to sink in the opinion of his friends."[7] In spite of these initial drawbacks, entomology was said to have generated considerable interest in Great Britain in the early decades of the nineteenth century, and Kirby and Spence's *Introduction* was an important contributor to this development. In 1826 an Entomological Club was founded in London and seven years later the Entomological Society of London was established.[8] In the early part of the nineteenth century, entomology was relatively little studied in the United States. Thomas Say attributed the lack of interest to the size of the entomologist's object of investigation and to the difficulties of the study itself.[9]

Though Harris's 1839 statement to Doubleday gives a sense of how he thought he might be perceived, such an admission is not common for him. If he thought that his studies were undervalued, it was not from a sense that he was an object of other people's amusement. One source of Harris's usual confidence in the value of his studies undoubtedly came from his foundational belief in their practical usefulness, but he also was assured of the enjoyment that the study itself brought him and could bring others. When he dealt with the matter of public attitudes toward entomology, he did it in a constructive way and in relation to natural history in general. In fact, he addressed the image issue, on his own terms, in the introduction to the *Treatise*, writing, "some men of mean and contracted minds have made themselves merry at the expense of naturalists, and have sought to bring the writings of the latter into contempt, because of the scientific language and names they were obliged to employ." Having centered the problem of study in the use of language, he then admitted that entomology was uncommonly characterized by such difficulties.[10] His argument, therefore, offsets the source of difficulty from the natural object of study itself, to the human problem of how to contend with the myriad forms and varieties.

None of this can answer the question of whether Harris considered his life's work a success or not. There is no end-of-life reflection on Harris's part, but only those statements in the midst of living, and which were subject, to some degree, to the situation at the moment. Among the biographer's haunting questions is whether Harris felt particular pain in the fact that he never consummated the lepidopterous studies that he envisioned and to which, for a time, had devoted so much effort, and that events of his life were such that he became, after 1842, more and more a practical entomologist. A letter written to Ebenezer Emmons in 1845 comes closest to revealing this dark side of despair, his usual altruism, present elsewhere even in the absence of optimism, giving way to concern for personal reputation and a tinge of jealousy for those who will find greater opportunity than he. It is a stark self-revelation to a man— Emmons—whom Harris apparently did not know, and who had opened the wounds through a suggestion that Harris might want to contribute articles on insects to his *American Quarterly Journal of Agriculture and Science*. As he had to others, Harris referred to the burden of his duties in the library and the forced abeyance of his work in natural history, and this after twenty-five years of preparation and accumulation of entomological knowledge and resources which were directed toward "the design of preparing & publishing a work which should supply the place of a long wanted manual upon our insects." Harris then went deeper into his inner-self and confessed that

> to find the time [had been] spent in vain, as to any honor & profit to myself or advantage to science, requires an extent of philosophical indifference that I have not yet schooled myself into. It is but little satisfaction to find that the poor & confessedly imperfect treatise prepared in the hurry of other pressing engagements, to answer the calls of the State Legislature, & never designed to take the place of the more scientific & extensive work contemplated, has met with some favor—so much, at least, as to furnish many articles for various other publications.—And however well it may be for the cause of the science, that other laborers have appeared in the field I have been compelled to abandon.—Neither the one nor the other are sufficient to relieve me from disappointment or to compensate me for my lost labors.[11]

Even this statement of extreme bitterness and disappointment, however, shows the constant factor in his life—the study of insects was its central concern. Though Harris was less than sympathetic to those who were entering the field, as suggested in his letter to Emmons, his sustained actions spoke more than his words. The time and effort he spent, in later years, in a mentoring relationship with the representatives of entomology's two branches—John Lawrence LeConte in systematics and Asa Fitch in applied studies—is the best evidence of his faith in the interest and importance of studying the insect world.[12]

RECOGNITION IN HIS LIFETIME AND THE FACTOR OF PERSONALITY

While Harris's self-assessment is difficult to fathom, there is evidence of how he was perceived by his contemporaries. Instances were referred to in earlier chapters, including his election to membership in scientific societies. His friend Nicholas Hentz was at the Academy of Natural Sciences of Philadelphia when Harris was nominated for membership in 1826 and reported that "They rejoiced at the chance of obtaining so useful an addition."[13] Harris's *Treatise* was universally praised among his contemporaries, as indicated in earlier discussions, including published reviews noted in chapter 3. His friend Edward Doubleday reported that in England "We are all delighted with your Report."[14] Whether the *Treatise* was a decisive factor in his election as corresponding member of the Entomological Society of London is uncertain. When John Obadiah Westwood, a founder and secretary of the society,[15] wrote to Harris in 1845 to announce his election, he mentioned that he had not received the promised copy of Harris's *Treatise*.[16] Harris and Westwood had been occasional correspondents since 1835, so Westwood knew of him. Whether or not Westwood had actually seen the *Treatise*, it is likely that Doubleday would have mentioned its virtues, since the two British entomologists began publication of their *Genera of Diurnal Lepidoptera* in 1846.[17] In the letter announcing his election, Westwood thanked Harris for his paper on a new insect (a Goliath beetle).[18] The paper had been sent to him by Thomas S. Savage and Westwood noted that, at Savage's request, he had published a colored illustration of the insect in his *Arcana Entomologica; or, Illustrations of New, Rare, and Interesting Insects* (London, 1841–45).[19] Two years later, Savage wrote to Harris in regard to Westwood's publication on another Goliath beetle that Savage had collected in Africa and quoted Westwood as saying, "I have described and figd the magnificent new Goliathus which you forwarded . . . We have given it the name of Harrisii as you desired, in honor of a man in every way worthy of having so splendid an insect named after him. It forms a distinct subgenus &

will be published in the Trans. Ent. Soc."[20] It is clear, therefore, that Westwood held Harris in high regard as an entomologist.

In the United States, Harris received praise from various quarters, in addition to the major reviews of his work. In 1844, a Boston correspondent wrote to thank him for the volume and in his comments inferred a high place for Harris's writings among the products of the state survey, offering the opinion that "Some of the mis-named scientific reports, belonging to the series brought out under legislative [illegible word], shall not slide quietly into oblivion, without passing through an alembic that will make their begetters wince— There is a vast difference between words and ideas—notwithstanding the fact that many take it for granted that both mean the same thing."[21] Planter and agricultural writer Thomas Affleck in Mississippi, as indicated in an earlier chapter, was one of those who appreciated the value and impact of Harris's writings on American entomological interest during his lifetime. Indeed, Harris's reputation among those interested in insect damages elevated him to a central resource. As a Philadelphia correspondent put it, "You must excuse this troubling you, but you are a recorder of entomological deeds for the United States, and it is a duty with all those who take an interest in Natural History or in you, to send their observations though perhaps, as in the present case, vague, imperfect and superficial."[22] Though interest in the *Treatise* appears to have been widespread, Harris thought at one time that it "has been much more praised than read."[23]

Judged by the content of his published writings and correspondence, Harris was able to deal with detailed information that required many hours of application to evaluate and assimilate. His writing style is straightforward, clear, factual, without pretense or conscious artistic quality. It was, in effect, well suited to its purpose and seems a true representation of the man himself. The characterizations of Harris by earlier biographers, some of whom knew him, tend to confirm the reading of his personality as revealed in the tone and content of his own writings, though other dimensions are added. An obituary, probably written by his classmate John G. Palfrey, said of him when he was in college that he had "a reputation for good scholarship and irreproachable morals, and with the esteem of all who had come into relations with him, though a certain constitutional shyness and self-distrust had prevented him from forming intimacies to the extent usual at that period of life."[24] His son, however, did not think that the undergraduate qualities of shyness and sensitivity applied to his father's adult years. Among his qualities, what impressed the son was Harris's capacity for labor.

His powers for work seemed exhaustless. He apparently needed no rest. His life was one of untiring activity and of constant occupation. No man ever knew better the value of time, or how best to economize it. All his efforts were directed to some well-defined purpose, and this was undoubtedly the reason that he was able to accomplish in his lifetime such an amount of varied and useful work. . . . While not possessing a buoyant and ardent temperament, he was ever cheerful and composed. If in occasional private letters to those who were in close sympathy with him in his scientific pursuits he expressed grief and disappointment at his inability to devote his entire time to the studies that so largely engrossed his thoughts, he did not suffer that feeling to embitter his life or to cramp his energies for work. He was by nature a silent, reserved man, and, as he passed middle life, it is possible that he grew more absorbed in his own thoughts and less open to approach than in his younger days.

The son's characterization continues: "Of all men he was one of the most simple and unostentatious in his tastes, habits of life, dress, and manners. He had an intense hatred of all shams and deceit. In conversation he was simple and unaffected, simple in thought and expression, and thoroughly earnest and sincere in all his doings."[25]

Harris's daughter, who was born in 1844, characterized her father in terms of his generosity. The recipients included students, and family members who apparently took advantage of his charitable qualities. "Many a stranger, seeking advice or help, sought Father at the Library and went on his way encouraged, wisely counselled and not infrequently pecuniarily helped. Far up in the north of Norway, a fellow-traveller on the steamer, finding that I was Dr. Harris' daughter, exclaimed with emotion, 'Oh, never in all my life was any one so kind to me as your father! I was a stranger in a strange land, poor and homesick, and he was so good to me!' "[26] Harris's biographers have noted that the same generous qualities characterized his role in the entomological community, helping younger entomologists, other naturalists, and farmers.[27] In relation to the latter, his son wrote that

After the publication of his "Treatise," and in fact until his life was ended, he was the recipient of constant calls and communications from agriculturists and others seeking information concerning habits of destructive insects and the proper remedies for their extinction; and many an old farmer who had travelled miles to bring him a newly discovered cabbage-pest, or a strange wheat-fly, was sent away from his house delighted with the story of insect life, and the practical hints that he received from the reserved but courteous gentleman who had welcomed him with the dignity and politeness of a by-gone age.[28]

Absorption in his work, modesty, generosity, the inclination to avoid controversy and confrontation,[29] but also feeling the pain of circumstances that stymied the course he envisioned and hoped for himself, these were facets that made him a good and effective entomologist and a moral voice in the community. They were part of a make up that was focused on a single branch of natural history, stretched between practical social needs and scientific study as an end in itself. As a totality, it seems likely that these qualities and preoccupations would also have deterred achievement of a high order as a leader of the naturalist community at large, even if his entomological career had not been curtailed by the events in his life in the early 1840s.

REPUTATION AT DEATH

Obituaries and resolutions by learned societies at the time of death have to be treated with considerable circumspection as sources of appraisal, but if approached with skepticism they have value for the things that are emphasized and to some degree the phrasing that is used. Harris's obituaries in the Boston papers reported that he was "favorably known throughout the scientific world for his investigations and publications in relation to insects, particularly those indigenous to Massachusetts,"[30] and "was a man of extensive scientific acquirements, and greatly distinguished himself as a Naturalist."[31] Interestingly, these accounts do not mention the relations of his entomological work to agriculture but identify him as a scientist and naturalist in a general sense. Harris's Harvard classmate John Gorham Palfrey gave a more extended newspaper account of his life, parts of which have been quoted in several other sections of this work. Palfrey praised his friend's character and his work but also recognized weaknesses and Harris's limited success in reaching the goals toward which he aspired. Of Harris's qualities and reputation as a scientist, Palfrey wrote

> He possessed those rare powers of observation, discrimination and analysis, which, united to a hearty love of the pursuit, make a naturalist of the highest order. He was a learned botanist; but the department of Natural History to which he was especially devoted was the study of the insect tribes. There appears to be no doubt of his being recognized, by the common consent of European naturalists, as the first entomologist in the world. Alike his modesty and his scanty means prevented his coming before the public with the results of his life-long study in a form becoming their curiosity and importance. His "Treatise on some of the Insects

of New England, which are Injurious to Vegetation," first published in 1841, under a commission from the Commonwealth, is a permanent contribution to science of the highest value. Published in so unexpensive a style as to be without even a single engraving, the descriptions are so perfect in their perspicuity and liveliness, that the mind obtains the image without assistance from the eye. The style of writing for compositions of this class cannot be excelled; but it is not attainable without a thorough mastery of the subject, united to a genuine love of it. The amount of information concerning the habits of the ephemeral life described, excites nothing short of a feeling of wonder at the sagacity and diligence that had amassed it.[32]

Palfrey recognized that the *Treatise* was not the end of Harris's work, except that circumstances made it so.

Resolutions were passed by several Boston-area societies on Harris's death, namely the Massachusetts Horticultural Society,[33] the American Academy of Arts and Sciences, and the Boston Society of Natural History. At the American Academy, the resolution was introduced by conchologist Augustus Addison Gould, which included this characterization:

[A]s a naturalist he has not been surpassed by any of his countrymen, and has exhibited a patience, thoroughness, and accuracy of observation in the various departments of Natural History, a truthfulness in the delineations both of his pencil and his pen, and a singular facility in employing language intelligible to the common reader and at the same time fulfilling all the requirements of science, which render him a model for the interrogator of Nature; and that, through a long life of untiring industry, he has accumulated and published a mass of original observations, of an eminently practical bearing, which have won for him high consideration both at home and abroad, and will constitute for him an enduring monument.

Resolved, That while both the scientific and the practical world are largely indebted to him for his published papers, it is to be regretted that very many others of equal importance, which are known to have been prepared, or are in process of preparation, remain unpublished; and that the Academy tenders its assistance in their publication.

The resolution of the Academy includes the fine balance of scientific and practical work that characterized Harris's life. Louis Agassiz seconded the resolution and added the observation (as summarized by the reporter) that "Harris had had few equals, even if the past were included in the comparison."[34] Agassiz's appraisal was not impulsive. Henry David Thoreau recorded in his journal for January 1, 1853 that during a visit to Cambridge shortly before that date, assistant librarian John Langdon Sibley had reported that Agassiz told him that Harris "was the greatest entomologist in the world" and had permitted Sibley to repeat his estimate.[35] Along the same lines, the American botanist, Edward Tuckerman, has been quoted on Harris, "Of other genuine naturalists I have read, but he is the only one I ever knew."[36] At the Boston Society, it was Jeffries Wyman who introduced the resolution on Harris's death. In his remembrance, Wyman noted that Harris was among the few entomologists who were able to combine a wealth of knowledge of systematics with an interest and familiarity with the habits of insects.[37] The Essex Institute put Harris's reputation in historical perspective and pointed out that, following the death of Thomas Say, Harris came to hold first rank among American entomologists.[38]

In the decade and a half after Harris's death, his colleagues undertook to consolidate his work and thus to preserve and promote his reputation. The 1862 illustrated edition of his *Treatise*, "enlarged and improved, with additions from the author's manuscripts and original notes,"[39] became the standard and historic edition. It was the product of a resolve of the state legislature, the editorship of the secretary of the State Board of Agriculture, and the supervision of the illustrations by Louis Agassiz. In addition, several of the most eminent American entomologists of the time contributed notes for various sections, namely John Lawrence LeConte on the Coleoptera, Philip R. Uhler on the Orthoptera and Hemiptera, John G. Morris on the Lepidoptera, Edward Norton on the Hymenoptera, and Carl Robert Osten Sacken (then with the Russian Legation at Washington) on the Diptera. These naturalists took the assignment seriously, a tribute to the esteem in which Harris was held, but in consequence some of their editorial comments are corrective and critical of his work.

About 1863, William Sharswood[40] and J. B. Lippincott & Company of Philadelphia issued a prospectus for *The Entomological Writings of Thaddeus William Harris*. It was to be edited by Sharswood with an estimated 450 pages and would include a portrait and illustrations. The intended work was to follow the typographical form of the state's new edition of the *Treatise*, reprinting his scattered papers from various journals, many of which were named in the circular. The announcement asked for the support of all who desired to make "this mass of original observations, of an eminently practical and scientific character" more available.[41] No evidence has been found that this publication ever appeared, but some aspects of the project were realized in 1869 with the Boston Society of Natural History's first Occasional Papers volume, the *Entomological Correspon-*

dence of Thaddeus William Harris, M.D., edited by Samuel H. Scudder. The project was enabled by the Society's possession of Harris's manuscripts and collection, by his practice of retaining copies of his letters, and by the aid and cooperation of several of his correspondents, some of whom made available letters for the project. In addition to the entomological portions of his letters, the volume reproduced Harris's "Contributions to Entomology," which had appeared in the *New England Farmer* 1828–29, other "articles or fragments, hitherto obscure or inaccessible" (from the agricultural press), and some of his manuscript notes.[42] It was not the systematic reprinting of his scattered publications that Sharswood envisioned, but in certain ways was something more important, the publication of information that would otherwise never have been public. It was an addition to Harris's corpus.

Both posthumous publications of Harris's work are significant reflections of the place that he occupied in American entomology in the years immediately after his death, as well, undoubtedly, of the needs of students of entomology in the United States at the time. The publication of the *Entomological Correspondence* in 1869 was the occasion for Thomas Wentworth Higginson's memoir, which drew on his own remembrances of Harris (especially when Higginson was a student at Harvard) as well as use of Harris's manuscripts. Higginson's sketch became the basis for much that was subsequently written about Harris. It reflected not only the substance of his work but also the qualities of the man, and in certain ways the two were presented as integral to an understanding and appreciation of the historical legacy. Thirteen years after Harris's death, Higginson wrote of the "steady growth of Dr. Harris's reputation," which was

> not due alone to his position as pioneer in American science during its barest period. It has grown because he proves to have united qualities that are rare in any period. He combined a fidelity that never shrank from the most laborious details with an intellectual activity that always looked beyond details to principles. No series of observations made by him ever needed revision or verification by another; and yet his mind always looked instinctively towards classification and generalization. He had also those scientific qualities which are moral qualities as well; he had the modesty and unselfishness of science, and he had what may be called its chivalry.[43]

There is nostalgia in Higginson's characterization that also appears in accounts by others writing on Harris—it is a hint of the old fashioned in his mode of science. In such references, Harris's immersion in and singular devotion to his studies resulted in a retrospective view of him that focused not only on his scientific contributions but on the character of the scientist as well. There was a danger for his long-term reputation that the element of personality that was seen in his work, and the trend toward experimental or morphological work in biology later in the century, would relegate him to the status of an antiquated figure, at best considered a type for the older naturalist, at worst, forgotten. The rise of practical entomology in the second half of the nineteenth century prevented that eclipse and in fact brought him the status of a founding father.

LONG-TERM REPUTATION

The continuing growth of Harris's scientific reputation, noted by Higginson in 1869, was a phenomenon that others referred to as the century progressed. By 1889, as mentioned in the previous chapter, Samuel H. Scudder said unequivocally, "it was entirely through his familiarity with the early stages of insects that he gained his preeminence," while recognizing that Harris's work overall was the basis of entomological study in the United States.[44] Augustus Radcliffe Grote, in celebrating what he characterized as the semicentennial of Harris's *Treatise,* in 1889, was impressed by the survival and development of the author's influence and the general appraisal of his work.[45] A decade later, Leland O. Howard, in a history of economic entomology in the United States, also noted that Harris's "scientific reputation has steadily grown."[46]

Grote made the point that Harris was not successor to the first notable American entomologist, Thomas Say, because Harris gave American entomology "a fresh turn, a useful impetus,"[47] meaning studies that aided agriculture. He viewed Harris as the originator of practical entomology in the United States, although he recommended his book to students of technical and economic insect studies alike. In fact, Grote recognized that Harris was a product of the period in which he lived, and in doing so revealed a political agenda for the development of entomology in his own time. While Grote realized that Harris's era had not been the time for separation of these two "departments" of entomology, by the late 1880s their union was no longer desirable and Grote "always deprecated the mixing up of technical entomology, in the reports of State entomologists, with economic entomology, their proper subjects."[48]

Economic entomology in the United States took a leadership role in the development of the field in Harris's later years and after his death,[49] and assessments of his historical place were tied to that development. Even so, at least some historical accounts do consider his entomological research in other than applied terms. Grote himself devoted a part of his paper to Harris's work on butterflies and moths. An expert in American Lepidoptera, Grote complimented Harris on the "rare excellence" of his technical writings on that order. Grote had a particular interest in Harris's writings on the histories of the Spinner moths (Bombyces), a group to which Harris had devoted some effort, and he gently criticized Harris's knowledge of names for the species in this group, commenting, "The natural impatience with which inexperienced entomologists, in their rage for exact nomenclature, are apt to feel at his occasional mistakes, leads them to neglect this portion of the report," which he recommended they read and contemplate.[50] Grote's copy of the treatise, interestingly, was one that Harris had presented to Asa Fitch[51] and included marginalia by Fitch and apparently also by Harris.[52] Fitch's notations as referred to by Grote concerned names of insects and differences of opinion about the habits of species. Grote also mentioned some mistakes in nomenclature by Harris not noted by Fitch and made the general point that Harris "not unfrequently but mistakenly, identifies allied Southern forms . . . with New England species."[53]

Leland O. Howard, who was associated with the U.S. Department of Agriculture's Bureau of Entomology from 1878 and was its head after 1894,[54] thought that Grote's account and characterization of Harris was the most aesthetically pleasing. But in a less overtly political way than Grote, Howard undertook to evaluate the nature of Harris's science and was more inclined to see a contradiction between the public intention of Harris's published work and the character of the man and his motivation. Although Howard plainly labeled Harris "the founder of applied entomology in this country,"[55] and pointed out that his very first and more than half of his subsequent published papers were economic in orientation,[56] nonetheless he argued that "Doctor Harris was not by taste an *economic* entomologist. He was a lover of nature and student of insects because they interested him."[57]

Howard compares entomologists in the second half of the nineteenth century in order to gauge merit, including a comparative assessment of Harris. Among John Curtis in England, and Harris and Asa Fitch in the United States, Howard considered Fitch to be the best field worker and closest to agricultural interests. He commented on the reputation that Harris had for personal charm, while Fitch was historically neglected in this vein. Harris also had the edge in priority and perhaps in literary style, as well as the fact that his work was accessible in a single volume and with the illustrations added in the posthumous edition. Howard suggested that Fitch also was overshadowed by Charles Valentine Riley, who began a rise to prominence in American economic entomology toward the end of Fitch's life and received a great deal more attention in his lifetime than did Fitch. Focusing his comparison on Fitch, Howard suggested that, in spite of the relative neglect, Fitch was as good an entomologist as either of the other two (Harris and Riley), and "there can be no doubt that many of his studies were as finished and as valuable as anything that Riley ever wrote and were more complete than anything that Harris did."[58]

Howard and other entomologists knew the value of Harris's *Treatise* through personal experience. Howard wrote in 1900, "On entering any entomological workshop in the land the first book that will catch the eye upon the desk is a well-worn copy of the 'Treatise upon insects injurious to vegetation.'"[59] He also wrote of his own delight as a fourteen-year-old when he received a copy of the third edition of Harris's *Treatise* for Christmas (his recollections seem to recall the illustrations in particular as a factor that drew his attention).[60] Entomologist John Henry Comstock's attestation of the value and effect of his copy of Harris's *Treatise* is noted prominently in a recent historical study of the influence of natural history texts in nineteenth-century America. Comstock wrote in his copy, in 1876, "I purchased this book for ten dollars in Buffalo, N.Y., July 2, 1870. I think it was the first Entomological work I ever saw. Before seeing it, I had never given Entomology a serious thought; from the time that I bought it I felt I should like to make the study of insects my life's work."[61] As with Harris's personal influence on agricultural entomologist Asa Fitch and systematist John Lawrence LeConte, his literary impression, through the *Treatise*, also influenced two later entomologists representing government and academic studies. Howard (1857–1950), who was a student of Comstock, served for many years as head of the Bureau of Entomology in the U.S. Department of Agriculture. Comstock (1849–1931) was professor of entomology and invertebrate zoology in Cornell University where, among other contributions, he taught some 5,000 students and published a number of important works.[62]

While Howard attests to the great practical influence of Harris's work,[63] others also have pointed out the ineffectiveness of his control measures in the era of big agriculture as it developed after the Civil War. The demonstrable success of the arsenical insecticide Paris green against the Colorado potato beetle in 1867 gave a boost to chemical usage, which came to dominate strategies for insect control. Nonetheless, Sorensen argues for the continued importance of knowing insect habits and relations (based on the idea of balance of nature) as a basis of all control,[64] a historical circumstance in which Harris's work could be valued not for its remedies but as a source of information on life histories and the identification of insects. In fact, Howard indicates that altering agricultural practices to counteract insect damage was an idea that had been rediscovered in the years preceding 1900, the time when he was writing, with the intimation that this development would give Harris's work on life histories a renewed currency.[65]

As noted above, the value and reputation of Harris's work did increase through the later decades of the nineteenth century. The *Treatise* was variously described as a standard or classical work, and one of continuing interest and value, and has been so designated up to recent times. In 1878, lepidopterist Herman Strecker advised that the 1862 illustrated edition was essential for all American entomologists.[66] In the works cited above, Scudder in 1889 called it an "acknowledged classic,"[67] and in the same year, Grote spoke for the continued reading of the *Treatise*.[68] The *Entomological News* from the Academy of Natural Sciences of Philadelphia in 1896 thought the 1862 edition remained an unrivalled introduction to its subject.[69] Historian Harry Weiss, writing in 1936, conceded that the remedies in the *Treatise* were of no continuing interest, but the book still was a classic in economic entomology,[70] and the next year Herbert Osborn urged that all entomologists should be familiar with it.[71] In 1971, historian of entomology Arnold Mallis still considered the *Treatise* to be a work of continuing value for its information on American insects.[72] It is apparent, therefore, that irrespective of the role of the *Treatise* in relation to Harris's own entomological program, or his self-concept of the direction in which he might have wanted his interests to go, the book by which he is known has in the century and a half since its publication had a life of its own, variously adapting to changing resources and needs among its readers. But it must be conceded, as well, that the stylistics that Harris gave it, and the grounding in fact that gave it a practical value, also elevated a substantial part of the content above the vagaries of scientific obsolescence.

Harris's wish that his collection should survive him and remain useful, even if his other projects were not fulfilled, was achieved,[73] although the passage of time inevitably took its toll. In 1941, his insects were transferred from the New England Museum of Natural History (Boston Society of Natural History) to the Museum of Comparative Zoology at Harvard University. The public announcement pointed out that it was likely the oldest general collection of American insects in existence and was in "fair condition considering its age." At the time, the curator made special note of the fact that the collection included types for at least two hundred insects that had been described by Say, Harris, and others.[74]

It is difficult to surmise what Harris would have thought about the public acknowledgments of his life and work. The inscription that Higginson composed for the plaque at the Suffolk Resolves house in Milton marking Harris's residence there mentioned his modesty,[75] a trait that was common in the estimates of the man, what the American Academy of Arts and Sciences' memorial resolution called his "eminent though unclaimed distinction."[76] Shortly after his death, in 1864, a Harris Entomological Club was established as a section of the Boston Society of Natural History and lasted until 1886. In 1899, a group of amateur entomologists organized in Boston as the Harris Club, which, in 1903 merged with the Cambridge Entomological Club.[77] Overshadowing these forms of remembrance, and in many ways epitomizing the historical reputation as well as the iconographic role that Harris's life and work came to play for American entomologists, was the presence of his portrait on the cover of the *Journal of Economic Entomology* (published by the American Association of Economic Entomologists) from 1917 to 1953.[78] The question of how Harris would react to this public view of his work as preeminently practical in nature lingers unanswered. It is true that, at one crucial span of years in his life, he put considerable effort into description, nomenclature, and classificatory relations among the Lepidoptera, and yet his life circumstances deferred that work beyond hope of completion. The entomology that he practiced, however, irrespective of the motivation or end view, was a broad-based one in which insect forms, actions, and relations were integral and lent itself to various perspectives or emphases, just as Harris himself stressed different purposes at different times in his life.

As a historical actor, aspects of Harris's life—his origins, education and occupational situation, his relationships with individuals and communities, and his underlying ideologies—were both controlled by personal actions and liable to contingencies, and all impacted the outcome. His entomological collecting and study filled his life with activity and accomplishment. As an intellectual figure, however, the physical product of his life—in particular, *A Treatise on Some of the Insects Injurious to Vegetation*—became both its symbol and its surrogate and once done, was no longer wholly his. Yet Harris's life also was used by the generations that followed—appeals to his personal character, for example, were used both to underscore the value and reliability of the *Treatise* and to bolster the ideal of the scientist.[79] For the biographer, the details and complexities of his life are a book by which to read the personal and historical situation in which he resided and acted and which gave being to its product. When the life was done, and the work broadcast, thereafter the man and his writing had a parallel existence, mutually reflective but also liable to the needs of their respective readers.

Note on Sources

LETTERS TO AND FROM HARRIS ARE LOCATED IN SEVERAL REPOSITORIES, and a number of these have been examined, as the references attest.* Most importantly, Harris often retained copies of his manuscripts, including drafts of many letters, and these are now in the Ernst Mayr Library at the Museum of Comparative Zoology, Harvard University. Matters relating to insects (largely of a technical character) in Harris's outgoing and incoming correspondence were extracted in the late 1860s and published as *Entomological Correspondence of Thaddeus William Harris, M.D.*, ed. Samuel H. Scudder, Boston Society of Natural History Occasional Papers 1 (Boston: Boston Society of Natural History, 1869). Although most of the personal element was deleted from the published letters, the volume is an excellent gauge and record of his scientific effort, when used in conjunction with the works he published during his lifetime. The endnotes to this book reference the original manuscript sources, but it should be kept in mind that sometimes a letter may have been extracted and published in the *Entomological Correspondence*, especially the portions dealing directly with insects. Sometimes the published version was used in lieu of the author's transcription but even in those instances the manuscript version is cited.

The *Entomological Correspondence* includes selections from correspondence with the following naturalists (number of letters in parentheses): Noyes Darling (1), Edward Doubleday (25), Nicholas M. Hentz (49), Edward C. Herrick (11), Thomas W. Higginson (1), William LeBaron (2), John L. LeConte (14), Frederick E. Melsheimer (5), Margaretta H. Morris (5), Thomas Say (1), and C. Zimmerman (1). Six letters from Harris to Thomas Say (with some of the drafts of Say's response) are in Samuel H. Scudder, [editor], "Some Old Correspondence between Harris, Say, and Pickering," *Psyche* 6 (1891–93): 57–60, 121–24, 137–41, 169–72, 185–87, 357–58 (being Harris letters dated July 7, 1823; December 22, 1823; November 18, 1824; February 21, 1825; May 15, 1825; and March 21, 1834); in the endnotes to this book, the original manuscript is cited.

A note in reference to Harris's retained copies of his letters is in order. These are, as indicated, generally drafts of letters rather than finished copybook versions. In some instances the letter as actually sent has been located, and there may be differences between it and the retained draft. Occasionally the differences are significant. It is important to keep in mind, therefore, that references or quotes from the drafts (which also appear in the published entomological correspondence) represent Harris's sentiments or attitudes, but were not necessarily part of a communication event between Harris and his correspondent.

*Harris's correspondence as Harvard University librarian is in the Harvard University Archives, but no systematic review of those records was undertaken for this study.

Abbreviations

Academy Correspondence, ANSP = Academy Official Correspondence, Academy of Natural Sciences of Philadelphia Library, Collection 567

Affleck = Thomas Affleck

Affleck Papers, LSU = Thomas Affleck Papers, Louisiana and Lower Mississippi Valley Collections, Louisiana State University Libraries

Agassiz = Louis Agassiz

Agassiz Papers, MCZ = Louis Agassiz Papers, Ernst Mayr Library of the Museum of Comparative Zoology, Harvard University, Ag 15.10.23

Am Acad Arts Scis Records = American Academy of Arts and Sciences. Records (manuscript minutes), American Academy of Arts and Sciences Archives, Cambridge, Mass.

ANSP = Academy of Natural Sciences of Philadelphia

APS = American Philosophical Society

Bigelow Papers, Houghton = John Prescott Bigelow Papers, Houghton Library, Harvard University, b MS Am 801.2 (775)

BSNH = Boston Society of Natural History

BSNH Minutes, MSciB = Boston Society of Natural History, Minutes of meetings, Museum of Science (Boston) Library

BSNH, Council minutes, MSciB = Boston Society of Natural History, Council minutes, Museum of Science (Boston) Library

Cambridge Scientific Club, General Folder, HUA = Cambridge Scientific Club, General Folder, Harvard University Archives, HUD 3257

CHS = Cambridge Historical Society, Cambridge, Mass.

Class of 1815 Memoranda, HUA = Harvard College, Class of 1815 Memoranda of the Class of 1815, Harvard University Archives, HUD 215.754

Class of 1815 Record Book, HUA = Harvard College, Class of 1815 Record Book, Harvard University Archives, HUD 215.776

Class of 1846 Class Book, HUA = Harvard College, Class of 1846 Class Book, Harvard University Archives, HUD 246.714

Correspondence between Say and Harris, MCZ = Correspondence between Thomas Say and Thaddeus William Harris and Others, with Papers and Notes, Ernst Mayr Library of the Museum of Comparative Zoology, Harvard University, MCZ F731

DeKay = James E. DeKay

Doubleday = Edward Doubleday

Everett Papers, MHS = Edward Everett Papers, Massachusetts Historical Society (microfilm)

First Parish Membership records, Andover-Harvard Library = First Parish (Cambridge, Mass.) Records: Membership records, Andover-Harvard Theological Library, Harvard University. I am grateful to former curator Timothy Driscoll for this information.

Fitch = Asa Fitch

Gould = Augustus A. Gould

Gould Papers, Houghton = Augustus Addison Gould Papers, Houghton Library, Harvard University, b MS Am 1210

Haldeman = Samuel S. Haldeman

Haldeman Correspondence, ANSP = Samuel Steman Haldeman Correspondence, Academy of Natural Sciences of Philadelphia Library, Collection 73

Harris Family Correspondence, HUA = Thaddeus William Harris Papers: Family Correspondence, Harvard University Archives, HUG 1445.405

Harris Family Papers, CHS = Harris Family Papers, Brinkler Library, Cambridge Historical Society, Cambridge, Mass.

Harris Letters to Say, Houghton = Thaddeus William Harris Letters to Thomas Say and Others, Houghton Library, Harvard University, MS Am 2093

Harris Papers, HU Bot Libs = Thaddeus William Harris Papers, Botany Libraries, University Herbaria, Harvard University

Harris Papers, MCZ = Thaddeus William Harris Papers, Ernst Mayr Library of the Museum of Comparative Zoology, Harvard University, bMu 1308.10 (except manuscript poem quoted in Chapter 5, which is call number bMu 1308.41.1)

Harris Papers [MSciB], MCZ, vol.x = Thaddeus William Harris Papers, Ernst Mayr Library of the Museum of Comparative Zoology, Harvard University, sf Mu 1308.43.x. This collection was formerly at the Museum of Science (Boston), where a microfilm copy has been retained. In individual citations, the "x" in the call number is replaced by the volume number. The letters by TWH in this collection are drafts or other retained copies.

Hentz = Nicholas M. Hentz

Herrick = Edward C. Herrick

Houghton = Houghton Library of Rare Books and Manuscripts, Harvard University

Houghton, Autograph File = Autograph File, Houghton Library, Harvard University

HU = Harvard University

HU Bot Libs = Botany Libraries, University Herbaria, Harvard University

HU College Papers, HUA = Harvard University, College Papers, Harvard University Archives, UA I 5.125 2nd series

HU Commencement program, HUA = Harvard University, Commencement program 1815 (printed), Harvard University Archives, HUC 6815

HU Corporation Papers (5.120), HUA = Harvard University, Corporation Papers, Harvard University Archives, UA I 5.120

HU Corporation Papers (5.130), HUA = Harvard University, Corporation Papers, Harvard University Archives, UA I 5.130

HU Corporation Records, HUA = Harvard University, Corporation Records (minutes), Harvard University Archives, UA I 5.30

HU Letters to the Treasurer, HUA = Harvard University Treasurer, Letters to the Treasurer, Harvard University Archives, UA I 50.8

HU Library Charging Records, HUA = Harvard University Library, Charging Records, Harvard University Archives, UA III 50.15.60

HU Library Letters, HUA = Harvard University Library, Letters (incoming correspondence), Harvard University Archives, UA III 50.8

HU Library Reports, HUA = Harvard University Library, Annual Reports of the Librarian, Harvard University Archives, UA III 50.5

HU Overseers Records, HUA = Harvard University, Overseers Records, Harvard University Archives, UA II 5.7

HU Overseers Reports (10.5), HUA = Harvard University Overseers, Reports, Harvard University Archives, UA II 10.5

HU Overseers Reports (10.6.2), HUA = Harvard University Overseers, Reports: Instruction Series, Harvard University Archives, UA II 10.6.2

HU Professorship of Natural History Records, HUA = Harvard University, Massachusetts Professorship of Natural History Records, Harvard University Archives, UA I 15.960 (pf and bound volume)

HU Professorship of Natural History Subscription Lists, HUA = Harvard University, Professorship of Natural History Subscription Lists, 1802–13, Harvard University Archives, UA I 15.997

HU Steward Quarter Bills, HUA = Harvard University Steward, Quarter Bills records, Harvard University Archives, UA I 70.15.100

HUA = Harvard University Archives

J.E. LeConte = John Eatton LeConte

J.L. LeConte = John Lawrence LeConte

LeConte Correspondence, ANSP = John Lawrence LeConte Correspondence, Academy of Natural Sciences of Philadelphia Library, Collection 913

Leonard = Levi W. Leonard

LSU = Louisiana and Lower Mississippi Valley Collections, Louisiana State University Libraries

MCZ = Ernst Mayr Library of the Museum of Comparative Zoology, Harvard University

Melsheimer = Frederick E. Melsheimer

MHS = Massachusetts Historical Society

J.G. Morris = John G. Morris

M.H. Morris = Margaretta Hare Morris

Mss Dept, APS = Manuscripts Department, American Philosophical Society Library

Mus Mss, MCZ = Museum Manuscripts, Ernst Mayr Library of the Museum of Comparative Zoology, Harvard University. Reference is to collections of manuscripts arranged by author and with call number preceded by Mu.

MSciB = Museum of Science (Boston) Library
Newhall to sister, HUA = Letter, Horatio Newhall to Lucy (sister), March
  3, 1815, Harvard University Archives, HUD 815.59
Palfrey Papers, Houghton = Palfrey Family Papers, Houghton Library,
  Harvard University, MS Am 1704 (419)
Peck Papers, HUA = William Dandridge Peck Papers, Harvard University
  Archives, HUG 1677.x
Pickering = Charles Pickering
Quinquennial folders, HUA = Quinquennial (biographical) folders, Har-
  vard University Archives, HUG 300
Say = Thomas Say
Sibley Journal, HUA = John Langdon Sibley, Private Journal, Harvard Uni-
  versity Archives, HUG 1791.72.10 (transcript available online at http://
  hul.harvard.edu/huarc/refshelf/Sibley.htm)
Sibley Library Journal, HUA = John Langdon Sibley, Library Journal, Har-
  vard University Archives, UA III 50.28.56.2
Sparks Papers, Houghton = Jared Sparks Papers, Houghton Library, Har-
  vard University, MS Sparks 153
Storer = David H. Storer
Storer Papers, MSciB = David H. Storer Papers (Letter Book), Museum of
  Science (Boston) Library (microfilm)
Sumner Papers, Houghton = Charles Sumner Papers, Houghton Library,
  Harvard University, b MS Am 1 (2922)
TWH = Thaddeus William Harris
Westwood = John Obadiah Westwood
Willcox = Edwin Willcox
Zimmerman (Zimmermann) = C. Zimmerman

Notes

PREFACE

1. Gillispie, *Professionalization of Science,* 3, argues, in fact, that Americans were largely absent from the rolls of major scientists even at the outset of the twentieth century.

2. Cravens and Marcus, "Introduction: Technical Knowledge in American Culture," 2.

3. Daniels, *American Science in the Age of Jackson,* 44.

4. Veysey, "Higher Education as a Profession," 22.

5. Guralnick, *Science and the Ante-Bellum American College,* 117.

6. The Rensselaer Polytechnic Institute was established in 1824 and the Sheffield Scientific School at Yale and the Lawrence Scientific School at Harvard were established in 1846 and 1847, respectively. Data in this chapter are taken from Elliott, *History of Science in the United States.*

7. Elliott, "Models of the American Scientist," 87.

8. James, *Elites in Conflict,* 29–30; Daniels, "Process of Professionalization in American Science," 157.

9. Porter, *Eagle's Nest,* 11, 136–37.

10. Kohlstedt, "Nineteenth-Century Amateur Tradition," 173–90 (quote, 178).

11. Kohlstedt, *Formation of the American Scientific Community,* 138–39; Holmfeld, "From Amateurs to Professionals in American Science," 22, 35.

12. Beaver, "American Scientific Community, 1800–1860," 16.

13. Porter, *Eagle's Nest,* 127–28. Porter puts Thomas Say and Thomas Nuttall in the category of excluded naturalists; both were important figures in the opening decades of the nineteenth century and in T. W. Harris's early scientific career.

14. Bruce, *Launching of Modern American Science,* 15, 44–45.

15. Guralnick, *Science and the Ante-Bellum American College,* ix.

16. The point is made, for example, by Cravens, "American Science Comes of Age," 54, and by Beardsley, *Rise of the American Chemical Profession, 1850–1900,* 47.

17. Guralnick, "American Scientist in Higher Education, 1820–1910," 127.

18. One study that defined the antebellum scientist in terms of journal article publication showed that about 40% of the authors were professors of science at some period in their lives. Elliott, "American Scientist, 1800–1863," based on data in Table 21 and p.117.

19. Merrill, *First One Hundred Years of American Geology,* 126.

20. Nash, "Conflict between Pure and Applied Science," 175; Hendrickson, "Nineteenth-Century State Geological Surveys," 138.

21. Beardsley, *Rise of the American Chemical Profession, 1850–1900,* 60–61.

22. Greene, *American Science in the Age of Jefferson,* 12; Fleming, *Meteorology in America,* xix.

23. Slotten, *Patronage, Practice, and the Culture of American Science,* 37, 73.

24. Beaver, "American Scientific Community, 1800–1860," 20; James, *Elites in Conflict*, 11.

25. Daniels, *American Science in the Age of Jackson*, 46.

26. See Zochert, "Science and the Common Man," 31, for a discussion of science in newspapers during that era.

27. Kohlstedt, *Formation of the American Scientific Community*, 116, 219–20.

28. Daniels, "Process of Professionalization in American Science," 163 n. 47; Kimball; *"True Professional Ideal" in America*, 209; Daniels, *American Science in the Age of Jackson*, 53.

29. Daniels, "Process of Professionalization in American Science,"166; Keeney, *Botanizers*, 100, 108.

30. Kimball, *"True Professional Ideal" in America*, 10.

31. Daniels, "Process of Professionalization in American Science."

32. Keeney, *Botanizers*, 6–7.

33. See Goldstein, "'Yours for science'," for a critique of studies that emphasize the professionalizing cadre of scientists and a report on a host of others who participated in science in the third quarter of the century.

34. Elliott, "Introduction: The Scientist in American Society." The typical scientist is based on the information in the tables in this work. (In the data source, the factor of multiple fields or occupations is dealt with by counting each individual as a fraction of one depending on the number of fields or occupations in which that person worked. Consequently, emphasis is on the ranking of the fields or occupations rather than the scientists directly.)

35. Beaver, "American Scientific Community, 1800–1860," 150–51; Bruce, *Launching of Modern American Science*, 94.

36. Lankford, *American Astronomy*, 14.

37. Of George Daniels' very select list of 56 scientists, 41 (73%) were professors of science. Daniels, *American Science in the Age of Jackson*, 32.

38. About a quarter of the scientists at one point in their lives did scientific work that was sponsored by government. See note 34.

39. Elliott, "American Scientist, 1800–1863," 136–40. (This citation relates only to the previous sentence.) The definition of a scientist in this source is based on journal article publication.

40. Elliott, "Models of the American Scientist," 86–87; Elliott, "Introduction: The Scientist in American Society," 4, 7.

41. The analysis is based on biological scientists who were contemporaries of T. W. Harris and included in Elliott, *Biographical Dictionary*, plus David Hosack, who was unfortunately omitted from that work. In the occupational total (using the fractional counting as described above [note 34]), medicine was 24.4% while professor of science was 17.5% of the occupational total, essentially the reverse of the percentages for the general scientific population.

42. Guralnick, *Science and the Ante-Bellum American College*, 109–10.

43. Subsequently privately reprinted as T. W. Harris, 1842, *Treatise on Some of the Insects*.

44. Unless otherwise indicated, information is based on Elliott, *Biographical Dictionary; American National Biography;* Sterling and others, *Biographical Dictionary of . . . Naturalists and Environmentalists*. (Peabody does not appear in any of these sources; Dewey and Emerson are not included in Sterling and others.)

45. Sterling and others, *Biographical Dictionary of . . . Naturalists and Environmentalists*, 251.

46. Peabody, *Sermons*, 20–21.

47. *Dictionary of American Biography.*

48. Cravens, "American Science Comes of Age," 55. Cravens uses this description in reference to the research achievement of antebellum professors who moved forward largely under their personal initiative, in the absence of a clearly defined place for research within the academic institutions. The personal factor applies at least equally to those committed to research but without the professorial connection.

49. Rosenberg, "Science in American Society," 365.

50. See, for example, Slotten, *Patronage, Practice, and the Culture of American Science,* 32, 145; Reingold, "Reflections on 200 Years of Science in the United States," 18–19.

ACKNOWLEDGMENTS

1. See the notes to the preface for some of these works. See also Elliott, "American Scientist in Antebellum Society," which includes comparisons to "men of letters" and inventors.

CHAPTER 1. ORIGINS

1. Hales, *Survey of Boston and Its Vicinity,* 48; Snow, *History of Boston,* 386; *History of the Town of Dorchester,* 386 (quote), 392.

2. *Family Story,* 55.

3. Frothingham, "Memoir of Rev. Thaddeus Mason Harris," 151.

4. *Concise Dictionary of American Biography.*

5. Higginson, "Memoir of Thaddeus William Harris," xii–xiii.

6. Frothingham, "Memoir of Rev. Thaddeus Mason Harris,"144. For a succinct account of the affair, see the entry for William Morgan (1774–1826?) in *American National Biography.*

7. Frothingham, "Memoir of Rev. Thaddeus Mason Harris," 140; Kelly and Burrage, *Dictionary of American Medical Biography,* 533.

8. Frothingham, "Memoir of Rev. Thaddeus Mason Harris," 142–43.

9. TWH to Storer, April 25, 1842, Storer Papers, MSciB, pp.306–8.

10. E. D. Harris, "Memoir of Thaddeus William Harris," 313.

11. Frothingham, "Memoir of Rev. Thaddeus Mason Harris," 136; E. D. Harris, "Memoir of Thaddeus William Harris," genealogical table.

12. Lincoln, *History of Worcester, Massachusetts,* 255 and 263 (the latter quoting an account by Thaddeus Mason Harris). Also see two recent biographies of Thaddeus William Harris's cousin: Gollaher, *Voice for the Mad,* and Brown, *Dorothea Dix,* which include useful information on Elijah Dix, his wife, and the Dix family.

13. Frothingham, "Memoir of Rev. Thaddeus Mason Harris," 138–39.

14. E. D. Harris, "Memoir of Thaddeus William Harris," 320.

15. Frothingham, "Memoir of Rev. Thaddeus Mason Harris," 131–32, 139.

16. See Sammarco, *Dorchester,* 36–37, and Sammarco, *Dorchester,* vol. 2:19 for photographs and notes on the house. The Harrises sold the house in 1840 and it was razed in 1916.

17. Frothingham, "Memoir of Rev. Thaddeus Mason Harris," 139. For general information on the family, see Mason, *Descendants of Capt. Hugh Mason,* 120–21.

18. TWH to Jared Sparks, August 18, 1842, Sparks Papers, Houghton.

19. TWH to Jared Sparks, [1848], Sparks Papers, Houghton.

20. Librarian's Report to the Examining Committee, July 11, 1842, HU Overseers Reports (10.5), HUA, vol. 6, and HU Library Reports, HUA.

21. Clarendon Harris to TWH, January 23, 1847, Harris Family Correspondence, HUA.

22. TWH to Jared Sparks, May 2, 1844, Sparks Papers, Houghton.

23. Class of 1815 Record Book, HUA.

24. E. D. Harris, "Memoir of Thaddeus William Harris," 314.

25. Shipton, *Biographical Sketches.*

26. TWH to Pickering, March 3, 1828, Harris Papers [MSciB], MCZ, vol.12.

27. Harvard University, *Catalogue of the Officers and Students* (broadsides, 1811, 1812, 1813, 1814); Quinquennial folders, HUA (biographical folder on Charles Briggs AB 1815).

28. Morison, *Three Centuries of Harvard,* 195–96.

29. Bailyn and others, *Glimpses of the Harvard Past,* 20.

30. Ibid., 23.

31. Harvard University, *Laws,* 32–33, and appendix p.4–6.

32. Hill, *Life at Harvard.*

33. HU Library Charging Records, HUA; Harvard University, *Laws,* appendix p.4.

34. Class of 1815 Record Book, HUA.

35. Newhall to sister, HUA.

36. HU Commencement program, HUA.

37. Class of 1815 Memoranda, HUA; this manuscript was the gift of Harris's daughter Elizabeth Harris to the library in 1924, and the entry for Harris (if not the entire booklet) almost certainly is in T. W. Harris's hand. Harvard University, *Catalogue of the Officers and Students,* issues published October 1818, 1819, 1820.

38. Harvard University, *Statutes.*

39. *Family Story,* 35.

40. TWH to Doubleday, November 17, 1842, Harris Papers [MSciB], MCZ, vol.13; Class of 1815 Memoranda, HUA.

41. Higginson, "Memoir of Thaddeus William Harris," xiii.

42. Barrows, "Dorchester in the Last Hundred Years," 3:595.

43. Frothingham, "Memoir of Rev. Thaddeus Mason Harris," 144.

44. *History of the Massachusetts Horticultural Society,* 501.

45. Thaddeus Mason Harris to TWH, n.d. [ca.1837?], Harris Papers [MSciB], MCZ, vol.14.

46. TWH to Hentz, November 19, 1828, Harris Papers [MSciB], MCZ, vol.13; TWH to Hentz, December 19, 1828, Harris Papers [MSciB], MCZ, vol.13; Higginson, "Memoir of Thaddeus William Harris," xii and xiii.

47. Lincoln, *History of Worcester,* 255 and 263.

48. TWH to Say, July 7, 1823, Harris Papers [MSciB], MCZ, vol.14.

49. TWH to M. H. Morris, December 1852, Harris Papers, MCZ.

50. Higginson, "Memoir of Thaddeus William Harris," xiii.

51. TWH to C. J. Schonherr [Skara, Sweden], December 30, 1836, Harris Papers [MSciB], MCZ, vol.15; Higginson, "Memoir of Thaddeus William Harris," xiii.

52. Howard, "Harris, Thaddeus William"; Dow, "Work and Times of Dr. Harris," 106; Graustein, "Natural History at Harvard College," 78.

53. Harvard University, *Laws,* appendix 4–5.

54. The evidence is largely of a negative character, in effect, the absence of cor-

roboration of enrollment. There is no evidence of payment of the required fee in HU Steward Quarter Bills, HUA, for the period from the fourth quarter 1813–14 through the fourth quarter 1814–15. HU, *Laws,* appendix p-5, notes that sons of subscribers to the fund for the Massachusetts Professorship of Natural History are not charged for the course, but Harris's father does not appear to have been a subscriber. See HU Professorship of Natural History Subscription Lists, HUA, and HU Professorship of Natural History Records, HUA.

55. Higginson, "Memoir of Thaddeus William Harris," xiii–xiv; Dow, "Work and Times of Dr. Harris," 106.

56. HU Library Charging Records, HUA, 1814–15.

57. TWH to Hentz, February 26, 1828, Harris Papers [MSciB], MCZ, vol.13; TWH to Pickering, March 3, 1828, Harris Papers [MSciB], MCZ, vol.12.

58. TWH to Storer, June 14, 1838, Storer Papers, MSciB, 296–98.

59. TWH to Doubleday, November 17, 1842, Harris Papers [MSciB], MCZ, vol.13; *Family Story,* 32.

60. E. D. Harris, "Memoir of Thaddeus William Harris," 314–15.

61. TWH to Storer, June 14, 1838, Storer Papers, MSciB, 296–98.

62. Howard, "Progress in Economic Entomology," 136.

63. Morris, "Contributions toward a History of Entomology in the United States," 20.

64. Essig, *History of Entomology,* 731; Morris, "Contributions toward a History of Entomology in the United States," 19.

65. Higginson, "Memoir of Thaddeus William Harris," xiii.

66. Smallwood, *Natural History and the American Mind,* 304.

67. Graustein, "Natural History at Harvard College," 78, 75.

68. TWH to Marshall P. Wilder, December 18, 1840, Harris Papers [MSciB], MCZ, vol.14, and *New England Farmer* 19, no. 51 (June 23, 1841): 405.

69. TWH to Storer, June 14, 1838, Storer Papers, MSciB, 296–98.

70. For Lowell and the Society, see Thornton, *Cultivating Gentlemen,* esp. 89–90, 174 (also see index).

71. TWH to Storer, June 14, 1838, Storer Papers, MSciB, 296–98.

72. Thaddeus Mason Harris to TWH, n.d. [ca.1824?], Harris Papers [MSciB], MCZ, vol.14. Peck's manuscript lectures, along with some correspondence and other papers, are in the Harvard University Archives, Peck Papers, HUA.

73. [Letter, probably TWH to John Lowell, addressed to "Respected & dear Sir"], n.d. [ca.1824?], Harris Papers [MSciB], MCZ, vol.14.

74. Weiss and Ziegler, *Thomas Say,* 110–11 (quote from TWH letter to Say, November 18, 1824).

75. E. D. Harris, "Memoir of Thaddeus William Harris," 317; Higginson, "Memoir of Thaddeus William Harris," xiv.

76. Higginson, "Memoir of Thaddeus William Harris," xx (quote from TWH letter to Storer, November 2, 1836).

77. T. W. Harris, 1823, "Upon the Natural History of the Salt-Marsh Caterpillar."

78. Howard, "Progress in Economic Entomology," 137.

79. Weiss, *Pioneer Century of American Entomology,* 105.

80. "List of the Writings of Thaddeus William Harris," xxxviii–xxxix, xlv; and bibliography compiled by the author.

81. Greene, *American Science in the Age of Jefferson,* 416.

82. See ibid., chapter 3, "Scientific Centers in New England," 60–90, for an overview of the sciences in the region during this period. The Linnaean Society is

noted on pp.74–75. Also important is Stone, "Role of the Learned Societies," Linnaean Society, 207–32. There is no indication that Harris had any role in the Linnaean Society's activities.

CHAPTER 2. CAREER AND FAMILY

1. TWH to Say, July 7, 1823, Harris Papers [MSciB], MCZ, vol.14.

2. TWH (New York City) to sister, Dorothea Harris, May 20, [1822], Harris Family Correspondence, HUA.

3. Greene, *American Science in the Age of Jefferson*, 100–101.

4. TWH to Say, February 21, 1825, Harris Papers [MSciB], MCZ, vol.14, which includes a brief undated note by Say.

5. Higginson, "Memoir of Thaddeus William Harris," xiv.

6. *Family Story*, 28, 35, 38; Class of 1815 Memoranda, HUA.

7. *Family Story*, 8–9; Morison and Commager, *Growth of the American Republic*, 1:179.

8. Field, "Harris Memorial Tablet."

9. *Family Story*, 40.

10. *Daily Advertiser*, Thursday January 24, 1856 (clipping), in Sibley, Collectanea Biographica Harvardiana.

11. TWH to Pickering, March 3, 1828, Harris Papers [MSciB], MCZ, vol.12.; TWH to Hentz, December 19, 1828, Harris Papers [MSciB], MCZ, vol.13.

12. TWH to Hentz, April 3, 1830, Harris Papers [MSciB], MCZ, vol.13.

13. Scudder, *Butterflies*, 1:656; *Family Story*, 40.

14. TWH to J. L. LeConte, December 21, 1852, LeConte Correspondence, ANSP; Bruce, *Launching of Modern American Science*, 135.

15. TWH to Hentz, February 6, 1826, Harris Papers [MSciB], MCZ, vol.13.

16. TWH to Hentz, July 27, 1827, Harris Papers [MSciB], MCZ, vol.13.

17. TWH to Leonard, February 12, 1828, Harris Papers [MSciB], MCZ, vol.14.

18. TWH to Pickering, March 3, 1828, Harris Papers [MSciB], MCZ, vol.12.

19. TWH to Hentz, December 19, 1828, Harris Papers [MSciB], MCZ, vol.13; Higginson, "Memoir of Thaddeus William Harris," xv.

20. Hentz to TWH, March 25, 1827, Harris Papers [MSciB], MCZ, vol.13.

21. Hentz to TWH, December 3, 1828, Harris Papers [MSciB], MCZ, vol.13; TWH to Hentz, December 19, 1828, Harris Papers [MSciB], MCZ, vol.13.

22. Hentz to TWH, January 1829, Harris Papers [MSciB], MCZ, vol.13; TWH to Hentz, January 16, 1829, Harris Papers [MSciB], MCZ, vol.13.

23. Hentz to TWH, March 1829 (and continued March 8), Harris Papers [MSciB], MCZ, vol.13.

24. TWH to Hentz, March 25, 1829, Harris Papers [MSciB], MCZ, vol.13.

25. Pickering to TWH, March 10, 1830, Harris Papers [MSciB], MCZ, vol.12.

26. TWH to Storer, June 14, 1838, Storer Papers, MSciB, 296–98; Dupree, *Asa Gray*, 106 (citing the foregoing letter). Relevant to the later discussion of Harris's aspirations for a professorship, and its relationship to his scientific work, is the particular situation that Peck occupied at Harvard. Robert McCaughey noted that Peck's relative freedom from teaching (requiring only a series of lectures to seniors) meant that a large part of his time was free for his research activities. In effect, the arrangement meant that Peck was able to see "himself as primarily a scientist and only incidentally a teacher." Peck was, in fact, "as close as anyone in Augustan Harvard [under the presidency of John Kirkland and Josiah Quincy] to being a pro-

fessional academic." McCaughey, "Transformation of American Academic Life," 254–55.

27. Graustein, *Thomas Nuttall,* 175–76; Morison, *Three Centuries of Harvard,* 217.

28. Graustein, *Thomas Nuttall,* 170–71, 175.

29. Ibid., 196–97.

30. TWH to Haldeman, November 15, 1842, Harris Papers [MSciB], MCZ, vol.14, and Haldeman Correspondence, ANSP; TWH to Doubleday, November 17, 1842, Harris Papers [MSciB], MCZ, vol.13.

31. Higginson, "Memoir of Thaddeus William Harris," xiii; TWH to Leonard, October 10, 1828 (continued on October 19), Harris Papers [MSciB], MCZ, vol.14. Also HU Corporation Papers (5.120), HUA, November 17, 1825 (appointed to committee); HU Corporation Papers (5.130), HUA, January 26, 1826 (on committee); HU Overseers Records, HUA, June 7, 1827 (appointed to fill vacancy).

32. Higginson, "Memoir of Thaddeus William Harris," xxxiii.

33. TWH to Hentz, June 5, 1829, Harris Papers [MSciB], MCZ, vol.13.

34. *Family Story,* 42–44, 47 and diagram opposite p.42; Doubleday to TWH, n.d. (received March 9, 1840), Harris Papers [MSciB], MCZ, vol.13.

35. *Family Story,* 43.

36. Librarian's Report to the President and Fellows of Harvard College, September 5, 1832, HU College Papers, HUA, vol.5.

37. *Family Story,* 42.

38. Higginson, "Memoir of Thaddeus William Harris," xvi.

39. *Concise Dictionary of American Biography.*

40. *Daily Advertiser,* Thursday January 24, 1856 (clipping), in Sibley, Collectanea Biographica Harvardiana.

41. TWH to John Lowell, January 25, 1827, Harris Papers, MCZ.

42. Pickering to TWH, March 4, 1832, Harris Papers [MSciB], MCZ, vol.12.

43. [TWH] to ? [the draft of a letter to George Bancroft is written between the lines of this letter and it is filed under his name], December 10, 1845, Harris Papers [MSciB], MCZ, vol.14; TWH to Haldeman, November 15, 1842, Harris Papers [MSciB], MCZ, vol.14, and Haldeman Correspondence, ANSP. Closer to the event, Harris cited a more personal reason for wishing the library appointment, namely the uncertain state of his health. TWH to Say, October 24, 1831, Harris Letters to Say, Houghton.

44. Concluding remarks, TWH, "Annual Address before the Boston Society of Natural History," May 1840, Harris Papers [MSciB], MCZ, vol. 6.

45. E. D. Harris, "Memoir of Thaddeus William Harris," 315. Edward Doubleday Harris was born in 1839 and therefore, to the degree that his comments are based on personal observation, would reflect especially the last decade or so of Harris's life. This is particularly relevant as concerns the time given to antiquarian studies.

46. TWH to Francis Boott, November 16, 1837, Harris Papers [MSciB], MCZ, vol.15.

47. Librarian's Report to the President and Fellows of Harvard College, August 24, 1836, HU College Papers, HUA, vol. 8, and HU Library Reports, HUA.

48. TWH to Francis [Boott?], February 14, 1842, Harris Papers [MSciB], MCZ vol.14. Also see TWH to Doubleday, April 13, 1840, Harris Papers [MSciB], MCZ, vol.13.

49. TWH to Zimmerman, November 10, 1837, Harris Papers [MSciB], MCZ, vol.15.

50. Hentz to TWH, September 8, 1839, Mus Mss, MCZ.

51. TWH to Dom O. J. Fahraeus, November 28, 1836, Harris Papers, MCZ.

52. TWH to Marshall P. Wilder, December 18, 1840, Harris Papers [MSciB], MCZ, vol.14, and *New England Farmer* 19, no. 51 (June 23, 1841): 405.

53. Higginson, "Memoir of Thaddeus William Harris," xvi.

54. TWH to Gould, December 30, 1834, Gould Papers, Houghton.

55. The reason for Boott's not accepting the post apparently was his inability or unwillingness to take responsibility for instruction in zoology in addition to botany. Dupree, *Asa Gray*, 107.

56. Elliott and Rossiter, *Science at Harvard University*, 340 (chronology).

57. For the latter see, e.g., TWH to DeKay, February 3, 1838, Harris Papers, MCZ. These engagements are discussed at length in later sections.

58. On May 16, 1838, David H. Storer introduced the topic of the Harvard professorship to the Boston Society of Natural History and a committee was consequently formed to explore the matter (BSNH Minutes, MSciB). On June 21, 1838, the Harvard Corporation referred a letter and treasurer's report on the professorship of natural history and botanic garden to a committee (HU Corporation Records, HUA, vol. 8: 49).

59. TWH to Storer, June 14, 1838, Storer Papers, MSciB, 296–98; TWH to Storer, July 4, 1838, Storer Papers, MSciB, 301.

60. TWH to Storer, June 14, 1838, Storer Papers, MSciB, 296–98.

61. Graustein, *Thomas Nuttall*, 321.

62. TWH to Gould, January 15, 1836, Gould Papers, Houghton.

63. TWH to DeKay, February 3, 1838, Harris Papers, MCZ.

64. TWH to Haldeman, November 15, 1842, Harris Papers [MSciB], MCZ, vol.14, and Haldeman Correspondence, ANSP. In 1842, at a time when his librarian's salary was $1,000, Harris received $274 for his zoological and botanical lectures and recitations. Receipt for teaching February 28–June 10, 1842, in Harris Papers [MSciB], MCZ, vol. 22.

65. TWH to Doubleday, April 13, 1840, Harris Papers [MSciB], MCZ, vol.13.

66. TWH to Francis [Boott?], February 14, 1842, Harris Papers [MSciB], MCZ, vol.14.

67. Graustein, *Thomas Nuttall*, 224; Graustein, "Natural History at Harvard College," 81.

68. Smallwood, *Natural History and the American Mind*, 251–52; Guralnick, *Science and the Ante-Bellum American College*, 111.

69. TWH to Doubleday, April 13, 1840, Harris Papers [MSciB], MCZ, vol.13.

70. For his lectures to sophomores, Harris used Henry McMurtrie's translation and abridgement of Cuvier's *Animal Kingdom*. See handwritten form for Harvard Department of Natural History, 1838–39, in Botanical Lectures, folder 2, Harris Papers, HU Bot Libs. This same note refers to the requirements of an "abstract of the lessons in writing" from the students. Instances of these abstracts are in the Harvard University Archives' curriculum collection for 1838; see there call numbers HUC 8838.358.x.

71. Dupree, *Asa Gray*, 106.

72. "Zoological Lectures. 1837," Harris Papers [MSciB], MCZ, vol. 7.

73. One page summary of zoology teaching, 1838–39, in Harris Papers [MSciB], MCZ, vol. 22.

74. Smallwood, *Natural History and the American Mind*, 308; Harvard University, *Annual Report* [1841–42], Report on "Department of Zoology and Botany."

75. Dupree, *Asa Gray*, 106.

76. Higginson, "Memoir of Thaddeus William Harris," xvii.

77. Smallwood, *Natural History and the American Mind,* 309.

78. Dupree, *Asa Gray,* 106.

79. TWH to Gould, January 15, 1836, Gould Papers, Houghton.

80. TWH to Storer, June 14, 1838, Storer Papers, MSciB, 296–98.

81. TWH to Say, March 19, 1829, Harris Papers [MSciB], MCZ, vol.14 (undated), and Mss Dept, APS.

82. TWH to Storer, June 14, 1838, Storer Papers, MSciB, 296–98. The date of the lectures is not known, nor has it been determined when the Dorchester lyceum was established. His papers include notes for lectures delivered (on entomological topics) to the Dorchester and Milton Lyceum in 1830 and 1831. Harris Papers [MSciB], MCZ, vol. 6.

83. TWH to Storer, June 14, 1838, Storer Papers, MSciB, 296–98; TWH to Storer, April 25, 1842, Storer Papers, MSciB, 306–8.

84. TWH to John Lowell, January 25, 1827, Harris Papers, MCZ.

85. TWH to Say, July 7, 1823, Harris Papers [MSciB], MCZ, vol.14.

86. TWH to John Lowell, January 25, 1827, Harris Papers, MCZ.

87. TWH to C. J. Schonherr, December 30, 1836, Harris Papers [MSciB], MCZ, vol.15.

88. TWH to B. W. Westermann, February 22, 1842, Harris Papers [MSciB], MCZ, vol.14; and Higginson, "Memoir of Thaddeus William Harris," xxiv (quoting that letter).

89. Boston Society of Natural History, *Proceedings* 7 (1859): 72.

90. T. W. Harris, 1869, *Entomological Correspondence.*

91. Scudder, *Butterflies,* vol. 1:657.

92. TWH to Lucy W. (Mrs. Thomas) Say, n.d. [ca. November/December 1836?], Harris Papers [MSciB], MCZ, vol.15.

93. TWH to Rev. Doct. [John Prince?], n.d., Harris Papers [MSciB], MCZ, vol.14. John Prince (1751–1836) almost certainly was the recipient. See Cohen, *Some Early Tools of American Science,* 63–64, 185 n. 49; Wheatland, *Apparatus of Science at Harvard,* 6, 133. Cohen (p.170 and appendix 3, illustration 44), and Wheatland (186–88) include illustrations of a lucernal microscope procured from Prince in the 1790s, an instrument for projecting an image; Prince is credited with certain improvements in the device.

94. TWH to Dom O. J. Fahraeus, n.d., Harris Papers, MCZ.

95. Bouvé, "Historical Sketch of the Boston Society of Natural History," 76; Higginson, "Memoir of Thaddeus William Harris," xxxv.

96. "Catalogue of the Harris Library."

97. TWH to J. S. Wright and J. A. Wight [*sic*], April 25, 1848, Harris Papers [MSciB], MCZ, vol.14.

98. TWH to Storer, November 2, 1836, Harris Papers [MSciB], MCZ, vol.15.

99. TWH to Michael Moore, Jr. (addressed to Trenton Falls, Oneida Co., NY), October 15, 1838, Harris Papers, MCZ.

100. TWH to Westwood, November 13, 1837, Harris Papers [MSciB], MCZ, vol.15.

101. TWH to J. E. LeConte, April 11, 1838, Harris Papers, MCZ, and Mss Dept, APS.

102. TWH to Messrs. Hector Bossange & Co. (Paris), October 8, 1840, Harris Papers, MCZ.

103. T. W. Harris, 1835, "Insects," Hitchcock *Report,* 601.

104. TWH to John Curtis, September 6, 1838, Harris Papers, MCZ (two nonidentical versions, a copybook and a draft). This relationship with Curtis is also consid-

ered in chapter 3 (regarding work on Lepidoptera) and chapter 4 (on sale and exchange of insects).

105. Higginson, "Memoir of Thaddeus William Harris," xviii–xix. Harris's papers, formerly in the Boston Museum of Science library, are now preserved in the library of the Museum of Comparative Zoology, Harvard University.

106. [TWH] to ? [the draft of a letter to George Bancroft is written between the lines of this letter and it is filed under his name], December 10, 1845, Harris Papers [MSciB], MCZ, vol.14.

107. Analysis of titles listed in "Catalogue of the Harris Library."

108. Hentz to TWH, October 8, 1826, Harris Papers [MSciB], MCZ, vol.13. No other reference to this resource has been noted.

109. TWH to Hentz, October 14, 1826, Harris Papers [MSciB], MCZ, vol.13; TWH to Pickering, November 19, 1826, Harris Papers [MSciB], MCZ, vol.12.

110. TWH to Hentz, November 3, 1827, Harris Papers [MSciB], MCZ, vol.13.

111. Meisel, *Bibliography of American Natural History,* 2:460, quoting from Gould, "Notice of the . . . Boston Society of Natural History," 236–41.

112. Librarian's Report to the President and Fellows of Harvard College, September 5, 1832, HU College Papers, HUA, vol.5.

113. TWH to President Josiah Quincy, June 24, 1834, HU College Papers, HUA, vol.6.

114. TWH to Gould, December 30, 1834, Gould Papers, Houghton.

115. Librarian's Report to the President and Fellows of Harvard College, September 13, 1838, HU Library Reports, HUA. As he noted to Say several years before, he willingly shared his personal books and specimens with other entomologists in the area. TWH to Say, June 14, 1834, Harris Letters to Say, Houghton.

116. Librarian's Report to the Examination Committee, July 13, 1855, HU Overseers Reports (10.6.2), HUA, vol. 1, and HU Library Reports, HUA.

117. TWH to Hentz, February 18, 1838, Harris Papers [MSciB], MCZ, vol.15.

118. TWH to Say, March 21, 1834, Harris Papers [MSciB], MCZ, vol.14. Although Harris sent the insects to Say in the fall of 1833, they were not received at New Harmony until March 1834. Say to TWH, April 1834, Correspondence between Say and Harris, MCZ. Say sent Harris the results of his examination throughout the summer; see letters from Say to Harris, June–August, 1834, Correspondence between Say and Harris, MCZ.

119. TWH to Pickering, November 13, 1834, Harris Papers [MSciB], MCZ, vol.12. Pickering resided in Philadelphia.

120. Lucy W. Say to TWH, March 28, 1835, printed in Weiss and Ziegler, *Thomas Say,* 214; TWH to Melsheimer, November 26, 1835, Harris Papers, MCZ.

121. TWH to Storer, November 2, 1836, Harris Papers [MSciB], MCZ, vol.15.

122. TWH to Lucy W. (Mrs. Thomas) Say, n.d. [ca. November/December 1836?], Harris Papers [MSciB], MCZ, vol.15.

123. TWH to Pickering, October 1836, Harris Papers [MSciB], MCZ, vol.15; TWH to Gould, July 9, 1836, Gould Papers, Houghton. In 1829, after his collection was shipped from Philadelphia to him at New Harmony, Indiana, Say reported that, on account of the condition of the insects at that time, he was compelled to dispose of two-thirds of the boxes, "preserving nothing but the pins;" Say to TWH, January 4, 1829, Correspondence between Say and Harris, MCZ.

124. TWH to Storer, November 2, 1836, Harris Papers [MSciB], MCZ, vol.15.

125. TWH to Willcox, January 25, 1838, Harris Papers [MSciB], MCZ, vol.15.

126. Bouvé, "Historical Sketch of the Boston Society of Natural History," 27; TWH to Edw. [H.?] Robbins, April 2, 1836, Harris Papers, MCZ.

127. TWH to Gould, February 9, 1836, Gould Papers, Houghton.

128. Harris "recycled" the printed letter he had prepared, using it as stationery; see TWH to J. E. LeConte, August 5, 1838, Harris Papers, MCZ, which is written on the back of one of the circular letters after the funds had been paid to Hentz.

129. TWH to Gould, n.d. [ca.1835], Gould Papers, Houghton.

130. TWH to Gould, February 19, 1836, Gould Papers, Houghton.

131. Bouvé, "Historical Sketch of the Boston Society of Natural History," 27.

132. Ibid.

133. TWH to Storer, November 2, 1836, Harris Papers [MSciB], MCZ, vol.15.

134. TWH to Hentz, December 1836 (note at top, "The substance of this letter sent March [14?], 1837"), and related letter April 24, 1837, Harris Papers [MSciB], MCZ, vol.15.

135. TWH to Prof. Ernst F. Germar (addressed to professor of mineralogy, Halle, Saxony), June 25, 1838, Harris Papers [MSciB], MCZ, vol.15.

136. Higginson, "Memoir of Thaddeus William Harris," xxvii.

137. E. D. Harris, "Memoir of Thaddeus William Harris," 316.

138. Higginson, "Memoir of Thaddeus William Harris," xxviii.

139. TWH to Doubleday, April 12, 1841, Harris Papers [MSciB], MCZ, vol.13.

140. Morison, *Three Centuries of Harvard*, 267.

141. Librarian's Report to the President and Fellows of Harvard College, September 13, 1838, HU Library Reports, HUA.

142. TWH to J. G. Morris, December [23 or 25], 1840, Harris Papers, MCZ.

143. TWH to Doubleday, April 12, 1841, Harris Papers [MSciB], MCZ, vol.13.

144. TWH to Dr. R. Bridges, Corresponding Secretary to Academy of Natural Sciences of Philadelphia, February 5, 1841, Academy Correspondence, ANSP, and printed in Weiss and Ziegler, *Thomas Say*, 208–9.

145. See TWH to Gould, July 9, 1836, Gould Papers, Houghton; TWH to Pickering, October 1836, Harris Papers [MSciB], MCZ, vol.15; Weiss and Ziegler, *Thomas Say*, 204, where the issue of date of transmittal is discussed with the same conclusion; TWH to Pickering, October 6, 1836, Harris Papers [MSciB], MCZ, vol.15.

146. TWH to Zimmerman, with letter to J. G. Morris, February 9, 1841, Harris Papers, MCZ.

147. Zimmermann to TWH, June 9, 1841, printed in Dow, "Work and Times of Dr. Harris," 111–13, and Weiss and Ziegler, *Thomas Say*, 207–8.

148. TWH to Walter R. Johnson, secretary to Academy of Natural Sciences of Philadelphia, January 26, 1842, Academy Correspondence, ANSP; Weiss and Ziegler, *Thomas Say*, 209–11. Also see Mawdsley, "Entomological Collection of Thomas Say," although his characterization of the Say Collection when in Harris's possession (p.164) is based on older published sources and is not supported by findings in the present study.

149. TWH to Melsheimer, January 4, 1842, Harris Papers [MSciB], MCZ, vol.12, and Harris Papers, MCZ; Melsheimer to TWH, February 14, 1842, Harris Papers [MSciB], MCZ, vol.12.

150. Meisel, *Bibliography of American Natural History*, 2:463.

151. Dupree, *Asa Gray*, 108–11 (quote, 109).

152. TWH to Francis [Boott?], February 14, 1842, Harris Papers [MSciB], MCZ, vol.14.

153. TWH to Storer, April 25, 1842, Storer Papers, MSciB, 306–8.

154. Frothingham, "Memoir of Rev. Thaddeus Mason Harris," 149.

155. TWH to Doubleday, November 17, 1842, Harris Papers [MSciB], MCZ, vol.13.

156. TWH to Storer, April 25, 1842, Storer Papers, MSciB, 306–8.

157. Dupree, *Asa Gray,* 112.

158. TWH to Doubleday, November 17, 1842, Harris Papers [MSciB], MCZ, vol.13.

159. Ibid.

160. The rumors that Harris had heard in April regarding the formation of a committee by the Boston Society, if true, must have had little or no relationship to the events taking place in Cambridge.

161. TWH to Doubleday, November 17, 1842, Harris Papers [MSciB], MCZ, vol.13.

162. Dupree, *Asa Gray,* 111–12.

163. Haldeman to TWH, October 31, 1842, Harris Papers [MSciB], MCZ, vol.14. Haldeman's address was published in a newspaper. See clipping, "(Reported for the Public Ledger) Introductory Lecture to a Course on Zoology, Delivered in the Hall of the Franklin Institute Nov. 17th, by S. S. Haldeman . . .", Harris Papers [MSciB], MCZ, vol. 22.

164. TWH to Haldeman, November 15, 1842, Harris Papers [MSciB], MCZ, vol.14, and Haldeman Correspondence, ANSP.

165. See, for example, TWH to Hentz, July 27, 1827, Harris Papers [MSciB], MCZ, vol.13, where he wrote that his letters are written during "any moment of leisure" over several days, and "then the rough copy (which is always kept for the purpose of reference) is transcribed without alteration."

166. TWH to J. G. Morris, February 5, 1845, Harris Papers, MCZ. See also TWH to Doubleday, January 30, 1844, Harris Papers [MSciB], MCZ, vol.13, and TWH to Herrick, March 24, 1845, Harris Papers [MSciB], MCZ, vol.12, where similar sentiments are expressed in summary form.

167. *Family Story,* 47 and 51; TWH to Doubleday, February 1844, Harris Papers [MSciB], MCZ, vol.13.

168. *Family Story,* 52.

169. The structure was 34 feet wide, but the length is not legible in Harris's letter.

170. TWH to Doubleday, February 1844, Harris Papers [MSciB], MCZ, vol.13.

171. *Cambridge Directory,* 1926 and 1930; Dow, "Work and Times of Dr. Harris," 107.

172. *Family Story,* 52.

173. Affleck (Ingleside, near Washington, MS) to TWH, September 8, 1846, Harris Papers [MSciB], MCZ, vol.14.

174. TWH to Affleck, November 29, 1849, as "Over 100 Years Ago . . ." (St. Louis, MO: Antimite Company, [1956]) (copy in: MCZ), and original letter in Affleck Papers, LSU (the "published" version by the Antimite Company is a handwritten transcription of TWH's original, with an introduction); *Concise Dictionary of American Biography.*

175. Affleck to TWH, September 8, 1846, Harris Papers [MSciB], MCZ, vol.14; Affleck to TWH, October 12, 1846, Harris Papers [MSciB], MCZ, vol.14. This inquiry occasioned some effort on Harris's part since the insect was not known to him. In 1848, Harris asked Affleck whether he would object if he published a "scientific description" of the insect in the *American Journal of Science* (TWH to Affleck, April 27, 1848, Affleck Papers, LSU), but no publication on the topic appears in Harris's bibliography.

176. Affleck to TWH, October 12, 1846, Harris Papers [MSciB], MCZ, vol.14.

177. Affleck to TWH, September 8, 1846, Harris Papers [MSciB], MCZ, vol.14.

178. Affleck to TWH, November 24, 1846, Harris Papers [MSciB], MCZ, vol.14 (quote); Affleck to TWH, January 7, 1847, Harris Papers [MSciB], MCZ, vol.14. Affleck saw the notice in the November 1846 issue of the *Farmer's Cabinet*, which was published in Philadelphia.

179. Affleck to TWH, January 23, 1850, Harris Papers [MSciB], MCZ, vol.14.

180. TWH to Affleck, February 11, 1850, Affleck Papers, LSU. This is the last letter that has been seen between the two men.

181. Doubleday to TWH, December 30, 1847, Harris Papers [MSciB], MCZ, vol.13.

182. TWH to Doubleday, March 24, 1849, Harris Papers [MSciB], MCZ, vol.13.

183. TWH to Westwood, [May?] [*sic*] 1854, Harris Papers, MCZ.

184. *Family Story*, 32.

185. Ibid., 24–25; E. D. Harris, "Memoir of Thaddeus William Harris," genealogical chart.

186. *Family Story*, 25–26 (quote, 26).

187. Ibid., 33–36 (quotes, 35).

188. Librarian's Report to the President and Fellows of Harvard College, August 24, 1836, HU College Papers, HUA, vol. 8, and HU Library Reports, HUA. Harris refers to "the depreciation in the value of money."

189. *Family Story*, 42; TWH to Edward Doubleday, April 13, 1840, Harris Papers, Harris Papers [MSciB], MCZ, vol.13.

190. *Family Story*, 55.

191. TWH to Samuel A. Eliot, January 4, 1848, HU College Papers, HUA, vol. 15. The "modal salary" of professors in 1829–30 was $1,500 and $1,800 in 1849–50. S. Harris, *Economics of Harvard*, 139.

192. TWH to Zimmerman, December 9, 1839, Harris Papers [MSciB], MCZ, vol.14.

193. This figure contradicts the statement in all of the published accounts on Harris, as well as some manuscript sources, where it is noted that he received only $175. A review of the expenses for the various surveys prepared in 1849 states that Harris received two payments of $175 (in 1838 and 1841). See Massachusetts, General Court—House, House Document no. 18 (1849). In 1852, Harris also affirms the higher amount, when he wrote that it was his recollection, "if I mistake not," that the commissioners each received about $300 [*sic*]; Brinckle, *Remarks on Entomology*, 8. The explanation for the differing accounts may be that, historically, reference to a payment of $175 for the *Report* was intended quite literally, that is, for the final report published as a monograph in 1841. The reference above to a payment in 1838 would have been for Harris's contribution to a volume of preliminary reports by the survey commissioners, T. W. Harris, 1838, "Report on the Habits of Some Insects Injurious to Vegetation." Whatever the exact amount, Harris has been called the first official or government-employed entomologist, based on the compensation he received for the state report. Howard, "Brief Account of . . . Official Economic Entomology," 58.

194. *Family Story*, 60.

195. Ibid., 38–40, 45, 56; Higginson, "Memoir of Thaddeus William Harris," xxxiv–xxxv.

196. TWH to Hentz, April 3, 1830, Harris Papers [MSciB], MCZ, vol.13.

197. TWH to Leonard, October 30, 1854, Harris Papers [MSciB], MCZ, vol.14.

198. TWH to Hentz, February 26, 1828, Harris Papers [MSciB], MCZ, vol.13.

199. TWH to Pickering, October 24, 1828, Harris Papers [MSciB], MCZ, vol.12.

200. TWH to Hentz, November 19, 1828, Harris Papers [MSciB], MCZ, vol.13.

201. E. D. Harris, "Memoir of Thaddeus William Harris," genealogical chart.
202. Ibid.
203. TWH to Gould, December 30, 1834, Gould Papers, Houghton.
204. TWH to ?, n.d. [November 1845?], Harris Papers [MSciB], MCZ, vol.14. The letter is undated but written on the back of one dated November 12, 1845. In the letter he refers to the daughter as age seventeen; if the letter is dated 1846, the daughter would be Harriet.
205. TWH to Affleck, September 18, 1848, Affleck Papers, LSU.
206. TWH to John L. Sibley, May 15, 1841, HU Library Letters, HUA, vol. 2.
207. Zimmermann to TWH, June 9, 1841, printed in Dow, "Work and Times of Dr. Harris," 111–13, and Weiss and Ziegler, *Thomas Say*, 207–8.
208. Zimmerman to TWH, July 4, 1853, Harris Papers [MSciB], MCZ, vol.14.
209. *Concise Dictionary of American Biography.*
210. *Family Story,* 49–50, 57.
211. Ibid., 55–56. Although this source is not identified as to author, it almost certainty was by Harris's daughter Elizabeth, who was born in 1844 and therefore only eleven years old when her father died. In addition to internal evidence, the copy of *A Family Story* with the Harris Family Papers at the Cambridge Historical Society has the handwritten note, "By Elizabeth Harris," on the title page.
212. TWH to Hentz, February 26, 1828, Harris Papers [MSciB], MCZ, vol.13.
213. *Family Story,* 60, 48 (quote), 65.
214. Class of 1846 Class Book, HUA, 227–34.
215. E. D. Harris, "Memoir of Thaddeus William Harris," genealogical chart; *Family Story,* 57–58, 61–62, 64; Walton, *Three-Hundredth Anniversary of the Harvard College Library,* 39; Mason, *Descendants of Capt. Hugh Mason,* 115–20; E. D. Harris, *New England Ancestors* (references to the situation of Harris's children in 1887 are derived from this work). Thomas Robinson Harris in *Report of the Secretary of the Class of 1863 of Harvard College,* 58–60. I am grateful to Brian Sullivan, who shared with me information on death dates of Harris family members, derived from the Cambridge City records.
216. TWH to Doubleday, March 24, 1849, Harris Papers [MSciB], MCZ, vol.13.
217. Dow, "Work and Times of Dr. Harris,"115; Osborn, *Fragments of Entomological History,* 145.
218. TWH to Hentz, November 19, 1828, Harris Papers [MSciB], MCZ, vol.13.
219. E. D. Harris, "Memoir of Thaddeus William Harris," 320.

CHAPTER 3. WORK IN SCIENCE AND IN THE LIBRARY

1. Classification of insects in Europe had three leading and successive figures: Carl Linnaeus (1707–78) whose system was based on the wings; Johann Christian Fabricius (1745–1808) who used the mouth parts as the most salient feature; and Pierre-André Latreille (1762–1833) who drew on multiple morphological features and promoted the idea of a more natural scheme of relationships. Weiss, *Pioneer Century of American Entomology,* 305; Lindroth, "Systematics Specializes," 122; Sorensen, *Brethren of the Net,* 2–4, 6.
2. This discussion is based on a visual scan and analysis of the citations listed in Meisel, *Bibliography of American Natural History* for the period 1768–1830.
3. Landon Carter, "Observations Concerning the Fly-Weevil that Destroys the Wheat; with Some Useful Discoveries and Conclusions, Concerning the Propagation and Progress of That Pernicious Insect, and the Methods to Be Used to Prevent

the Destruction of the Grain by It," *Transactions of the American Philosophical Society* 1 (1771 [read 1768]): 205–17; Howard, *History of Applied Entomology,* 11.

4. William Dandridge Peck, "The Description and History of the Canker-Worm," *Massachusetts Magazine; or Monthly Museum* 7 (1795): 323–27, 415–16; Elliott, *Biographical Dictionary.*

5. Citations in Meisel, *Bibliography of American Natural History;* Weiss, *Pioneer Century of American Entomology,* 58–85, 99–103. Sorensen, *Brethren of the Net,* 7, notes that early efforts in the United States relating to agricultural interests were less successful than descriptions of new species in earning international attention.

6. Elliott, *Biographical Dictionary.*

7. John Eatton LeConte, "Description of Some New Species of North American Insects," *Annals of the Lyceum of Natural History of New York* 1 (1824): 169–73; Weiss, *Pioneer Century of American Entomology,* 110. LeConte subsequently collaborated with French entomologists in producing works on American Lepidoptera and Coleoptera; Sorensen, *Brethren of the Net,* 10–11.

8. Stroud, *Thomas Say,* 45, 151–63, 226.

9. Sorensen, *Brethren of the Net,* 13–14; Mallis, *American Entomologists,* 25; Stroud, *Thomas Say,* 279; Weiss, *Pioneer Century of American Entomology,* 97. Mallis, Stroud, and Weiss make the point that Say's descriptive studies were a necessary prelude to further studies of life histories, etc.

10. In the case of journals, the total is only for those that included at least one entomological entry; that is, if a journal did not include papers on insects its other natural history articles are not counted. If the *American Journal of Science and Arts,* which appears to have published only four entomological articles from its founding in 1818 to 1830, and which overall was heavily weighted toward geology and mineralogy, is eliminated, the total number of natural history publications is about 1,350.

11. DeKay, *Anniversary Address on the Progress of the Natural Sciences in the United States,* 81, 86 (quote).

12. It has not been determined whether or not Harris was familiar with DeKay's essay.

13. Morris, "Contributions toward a History of Entomology in the United States," 17.

14. Sorensen, *Brethren of the Net,* 38, 253. Sorensen estimates that around 1870 about forty to fifty individuals "published regularly" in entomology while the larger group of writers on insects was about one hundred. Using a looser definition, he notes that self-identifying entomologists during the 1870s reached a total of about eight hundred. Morris, "Contributions toward a History of Entomology in the United States" notes eighteen individuals (deceased and living) who lived in the United States and who contributed to American entomology as of 1846.

15. Sorensen, *Brethren of the Net,* 60, 89–90.

16. Ibid., 64; Palladino, *Entomology, Ecology and Agriculture,* 24–27.

17. TWH to Say, July 7, 1823, Harris Papers [MSciB], MCZ, vol.14; Graustein, *Thomas Nuttall,* 188.

18. TWH to Say, July 7, 1823, Harris Papers [MSciB], MCZ, vol.14.

19. TWH to Pickering, October 24, 1828, Harris Papers [MSciB], MCZ, vol.12.

20. TWH to Hentz, February 26, 1828, Harris Papers [MSciB], MCZ, vol.13; TWH to Leonard, October 10, 1828 (continued on October 19), Harris Papers [MSciB], MCZ, vol.14.

21. TWH to Hentz, January 16, 1829, Harris Papers [MSciB], MCZ, vol.13.

22. Allen, *Naturalist in Britain,* 132–33.

23. Edward Doubleday in 1841 mentioned to Harris the success he and his

brother had had capturing Lepidoptera by brushing tree trunks with sugar and advised, "I think you would succeed in this way in getting a great many moths." Doubleday says they were following the method of Mr. Selby of Bedford, and does not indicate that he and his brother had pioneered in this method. Doubleday to TWH, October 19, 1841, Harris Papers [MSciB], MCZ, vol.13. On P. J. Selby and the method, see Allen, *Naturalist in Britain,* 132–33.

24. TWH to Say, November 18, 1824, Harris Papers [MSciB], MCZ, vol.14.

25. Although almost all of Harris's publications in this period are on entomology, he published one paper on a mole-like animal, in T. W. Harris, 1825, "Description of a Nondescript Species of the Genus Condylura."

26. For a modern review of this effort, see Hardy, "T.W. Harris and Cremastocheilus," where Hardy has compiled extracts from Harris's correspondence with Hentz and Haldeman on this group of insects.

27. TWH to Leonard, October 10, 1828 (continued on October 19), Harris Papers [MSciB], MCZ, vol.14.

28. See the introductory remarks to the first of the "Contributions," as reproduced in T. W. Harris, 1869, *Entomological Correspondence,* 337–38. The role of publication in the *Farmer,* in relation to his wider publishing program, will be taken up in a later section.

29. TWH to John Lowell, January 25, 1827, Harris Papers, MCZ.

30. TWH to Say, February 21, 1825, Harris Papers [MSciB], MCZ, vol.14, which includes a brief undated note by Say.

31. TWH to Hentz, May 16, 1825, Harris Papers [MSciB], MCZ, vol.13. Harris later reminded Hentz that he had told him about these plans as early as August 1824. TWH to Hentz, December 19, 1828, Harris Papers [MSciB], MCZ, vol.13.

32. TWH to Pickering, August 1826, Harris Papers [MSciB], MCZ, vol.12; Hentz to TWH, October 8, 1826, Harris Papers [MSciB], MCZ, vol.13.

33. Pickering to TWH, November 11 [1826], Harris Papers [MSciB], MCZ, vol.12.

34. TWH to Pickering, April 1826, Harris Papers [MSciB], MCZ, vol.12.

35. Pickering to TWH, April 26, 1826, Harris Papers [MSciB], MCZ, vol.12. It is not clear from this exchange between Harris and Pickering whether it was the Boston-area work or the more general cooperative project that was under discussion.

36. The full title of Bigelow's pioneering local work is *Florula Bostoniensis: A Collection of Plants of Boston and Its Environs, with Their Generic and Specific Characters, Synonyms, Descriptions, Places of Growth, and Time of Flowering, and Occasional Remarks* (Boston: Cummings and Hilliard; Cambridge: Hilliard and Metcalf, 1814). In 1824, Bigelow published a "greatly enlarged" edition, underscoring the popularity of the first.

37. TWH to Hentz, January 16, 1829, Harris Papers [MSciB], MCZ, vol.13.

38. TWH to J. E. LeConte, December 17, 1829, Harris Papers [MSciB], MCZ, vol.14, and Mss Dept, APS.

39. Scudder stated that many fragments of different parts of the Faunula were extant in Harris's papers then at the Boston Society of Natural History. At Scudder's prompting, the notes on butterflies were included with T. W. Harris, 1862, *Treatise on Some of the Insects* (the posthumous edition) and some of the notes on the preliminary stages of butterflies were reproduced in T. W. Harris, 1869, *Entomological Correspondence.* Scudder further observed, "This was the first tolerably complete descriptive list of the butterflies of any district in North America ever attempted." Scudder, *Butterflies,* 1:657.

40. TWH to Storer, November 2, 1836, Harris Papers [MSciB], MCZ, vol.15.

Among Harris's surviving papers are manuscripts titled "An Outline (Introduction) of (to) American Entomology for Children," which appears to date sometime after April 1833, and "Materials for a First Book on American Insects, for the Youth of New Eng. Also, Second Book on American Insects, an enlarged edition of the first. Jan. [2?] 1830." Harris Papers [MSciB], MCZ, vol. 7.

41. Dow, "Work and Times of Dr. Harris," 109. Dow makes the point that Harris's catalogue was a substitute for the never completed Faunula.

42. Hitchcock, *Report on the Geology, Mineralogy, Botany, and Zoology of Massachusetts*, 526.

43. T. W. Harris, 1833, "Insects," Hitchcock *Report*. The extent of Harris's effort is suggested by the fact that, of the 2,976 animal species listed in the first edition of Hitchcock's report, 2,350 were Harris's insects. T. W. Harris, 1835, "Insects," Hitchcock *Report*, 601; Meisel, *Bibliography of American Natural History*, 2:522.

44. T. W. Harris, 1833, "Insects," Hitchcock *Report*, 595. In his *Treatise*, he also used this formula, which he attributed to "an English entomologist," no doubt meaning William Kirby, but in a modified form based on the assumption that a proportion of 1 to 6 was too high for the United States, "where vast tracts are covered with forests, and the other original vegetable races still hold possession of the soil." Deciding on a ratio of 1 to 4, and still using 1,200 plants, in the *Treatise* he estimated the number of insects in Massachusetts as 4,800. T. W. Harris, 1842, *Treatise on Some of the Insects*, 10. The modern estimate of recorded insect species in the state, according to Leahy, *Introduction to Massachusetts Insects*, 4, is about 10,000.

45. T. W. Harris, 1835, "Insects," Hitchcock *Report.*, 601.

46. Ibid.

47. Weiss and Ziegler, *Thomas Say*, 176.

48. Higginson, "Memoir of Thaddeus William Harris," xxix.

49. TWH to John P. Bigelow, Massachusetts Secretary of State, February 11, 1840 (two letters, official and private), Harris Papers [MSciB], MCZ, vol.14, and Bigelow Papers, Houghton.

50. TWH to Pickering, March 25, 1830, Harris Papers [MSciB], MCZ, vol.12.

51. Farber identifies three ways in which the type concept was used in early nineteenth-century natural history. Harris's use was one that naturalists had resorted to for some time (though not necessarily in an openly avowed way), what Farber calls "classification type concept." In Harris's case, a particular species was used as a standard for its genus. Farber, "Type Concept in Zoology," 93–95.

52. Harris had initiated the correspondence with Westwood in September 1835 and sent insects. TWH to Westwood, September 25, 1835, Harris Papers, MCZ.

53. TWH to Westwood, June 1836, Harris Papers, MCZ.

54. See TWH to Zimmerman, November 10, 1837, Harris Papers [MSciB], MCZ, vol.15; TWH to Willcox, May 29, [1838?], Harris Papers, MCZ; TWH to B. W. Westermann, February 22, 1842, Harris Papers [MSciB], MCZ, vol.14.

55. Joseph D. Hooker to TWH, December 19, 1835, Mus Mss, MCZ.

56. Hooker, *Life and Letters*, 1:5, 25–26.

57. TWH to Joseph D. Hooker, June [6?], 1836, Harris Papers, MCZ.

58. Harris may be referring to T. W. Harris, 1832, *Discourse Delivered Before the Massachusetts Horticultural Society*. Among other subjects, Harris takes up the topic of insect and plant distribution in this lecture.

59. Joseph D. Hooker to TWH, July 31, 1837, Mus Mss, MCZ.

60. TWH to Haldeman, [January 9?], 1843, Haldeman Correspondence, ANSP. Harris addresses here what appears to be the question of what Michael Kinch calls "disjunct distributions" of species in a somewhat formulaic manner, by arguing

that lack of evidence of movement from one location to another meant that, what might have been the same species, really could not be the same, irrespective of how closely they resembled one another. Harris, of course, does not address the issue of what evidence he would consider for migration or introduction, but through omission seems to indicate that he was not willing to speculate on the matter. He certainly did not follow entomologist William Kirby who used "supernatural interventions and improbable migrations in order to explain the distribution of life." At the same time, although (according to Kinch) the question of distribution and the origins of species were closely linked in the early nineteenth century, Harris only assumes a separate origin of European and American species without any indication of the mechanism for such an event. See Kinch, "Geographical Distribution," 91–119 (quote on Kirby, 105).

61. "Botanical notes, by Thaddeus William Harris. On the Lycopsis Virginica of Bigelow" (MS), n.d.; also second version of the MS, June 4, 1842, with draft of letter, TWH to James E. Teschemacher, Harris Papers, HU Bot Libs; James E. Teschemacher to TWH, June 8, 1842, Harris Papers, HU Bot Libs.

62. TWH to Storer, November 2, 1836, Harris Papers [MSciB], MCZ, vol.15.

63. TWH to Westwood, June 1836, Harris Papers, MCZ .

64. TWH to Storer, November 2, 1836, Harris Papers [MSciB], MCZ, vol.15.

65. TWH to Melsheimer, November 26, 1835, Harris Papers, MCZ.

66. See Say, "Descriptions of New North American Insects," 155–190. This posthumous paper (a continuation of one Say had presented to the American Philosophical Society in 1832, and subsequently published), was read to the society on June 17, 1836. It has several bracketed editorial notations, one of which is identified as by "H" (p.181) and another by "T.W.H." (p.184), thus confirming that Harris had prepared it for publication. John L. LeConte, in his edition of Say, *Complete Writings*, notes that part of the paper (to p.168) had previously been published at New Harmony. The remainder (p.168–90) apparently was the portion that Harris prepared from Say's manuscripts. Presumably the publication of some of Harris's own insects was at stake in his efforts on behalf of Say's work; the portion of the paper in the "Descriptions of New North American Insects" (above) that Harris appears to have prepared from Say's manuscripts (i.e., 168–90) includes a number of references to insects provided to Say by Harris.

67. Say to TWH, July 19, August 7, and August 13, 1834; Lucy W. Say to TWH, October 15, 1834; and TWH to Lucy W. Say (draft), March 19, 1835, all in Correspondence between Say and Harris, MC3. Meisel, *Bibliography of American Natural History*, 2:471.

68. TWH to Lucy W. (Mrs. Thomas) Say, n.d. [ca. Nov./Dec. 1836?], Harris Papers [MSciB], MCZ, vol.15.

69. Weiss and Ziegler, *Thomas Say*, 216; TWH to J. G. Morris, September 17, 1839, Harris Papers, MCZ. See also Stroud, *Thomas Say*, 264–65; Stroud quotes from a letter from Harris to Maclure, December 8, 1836, in which Harris presents the case for Maclure's payment for the publication of Say's papers.

70. TWH to Zimmerman, November 10, 1837, Harris Papers [MSciB], MCZ, vol.15.

71. TWH to J. G. Morris, September 17, 1839, Harris Papers, MCZ. Harris must have been referring to Thomas Say, *Oeuvres entomologiques de Th. Say,* . . . recueillies et traduites par M. A. Gory (Paris: Lequien fils, 1837).

72. TWH to Joseph D. Hooker, June [6?], 1836, Harris Papers, MCZ. This characterization of his interests is not at all apparent in his entomological correspondence before this date.

73. TWH to Zimmerman, November 10, 1837, Harris Papers [MSciB], MCZ, vol.15.

74. TWH to Pickering, October 1836, Harris Papers [MSciB], MCZ, vol.15.

75. TWH to Zimmerman, November 10, 1837, Harris Papers [MSciB], MCZ, vol.15; Elliott, *Biographical Dictionary.*

76. TWH to Samuel G. Morton, March 13, 1838, Mss Dept, APS.

77. Samuel G. Morton to TWH, April 19, 1838, quoted in TWH to Samuel G. Morton, August 22, 1840, Academy Correspondence, ANSP.

78. TWH to Samuel G. Morton, July 16, 1838, Academy Correspondence, ANSP.

79. TWH to Samuel G. Morton, August 22, 1840, Academy Correspondence, ANSP. Morton's response to this letter has not been seen. Harris's friend, Zimmerman, reported that when he was in New York in September 1840, he had tried to bargain with Willcox to make the Townsend insects available as a gift to Harris but was not successful. Zimmermann to TWH, June 9, 1841, printed in Dow, "Work and Times of Dr. Harris," 111–13, and Weiss and Ziegler, *Thomas Say,* 207–8. While no direct evidence has been found, it is conceivable that the controversy over the Townsend collection and the question of ownership may have contributed to the Academy's decision to request the return of the Say collection several months later (see discussion in the previous chapter).

80. TWH to Pickering, January 17, 1838, Harris Papers [MSciB], MCZ, vol.15.

81. TWH to J. G. Morris, September 17, 1839, Harris Papers, MCZ.

82. TWH to Storer, November 2, 1836, Harris Papers [MSciB], MCZ, vol.15.

83. TWH to J. E. LeConte, August 5, 1838, Harris Papers, MCZ.

84. TWH to J. G. Morris, September 17, 1839, Harris Papers, MCZ.

85. TWH to Doubleday, May 8, 1839, Harris Papers [MSciB], MCZ, vol.13.

86. T. W. Harris, 1823, "Upon the Natural History of the Salt-Marsh Caterpillar."

87. TWH to Hentz, February 6, 1826, Harris Papers [MSciB], MCZ, vol.13.

88. TWH to Hentz, September 4, 1828, Harris Papers [MSciB], MCZ, vol.13.

89. TWH to Pickering, December 9, 1828, Harris Papers [MSciB], MCZ, vol.12.

90. Say to TWH, May 20, 1830 (quote) and November 28, 1830, Correspondence between Say and Harris, MC3.

91. Harris responded and declined the offer to take up the Lepidoptera, even though they "have always been very interesting to me," arguing the lack of access to European species. TWH to Say, October 15, 1830, MCZ, b MS 31.20.11.

92. No reference to the paper was found in the Am Acad Arts Scis Records.

93. Peale issued *Lepidoptera Americana: Prospectus* (Philadelphia, 1833), but his manuscript on "Butterflies of North America" (at the American Museum of Natural History) is unpublished. Elliott, *Biographical Dictionary.*

94. TWH to Say, August 13, 1834, Harris Letters to Say, Houghton; Say to TWH, August 13 and August 30, 1834, Correspondence between Say and Harris, MC3. After Say's death in October, his widow sent various materials to Harris but she was unable to find any reference to the promised Lepidoptera. Lucy W. Say to TWH, January 6, 1835, Correspondence between Say and Harris, MC3.

95. TWH to Westwood, September 25, 1835, Harris Papers, MCZ.

96. TWH to Storer, November 2, 1836, Harris Papers [MSciB], MCZ, vol.15.

97. TWH to Francis Boott, November 16, 1837, Harris Papers [MSciB], MCZ, vol.15.

98. TWH to Westwood, November 13, 1837, Harris Papers [MSciB], MCZ, vol.15.

99. T. W. Harris, 1869, *Entomological Correspondence*, 338. Harris's thoughts in regard to the *New England Farmer* and its relations to other, more scientific outlets will be discussed further in the next chapter.

100. TWH to Francis Boott, November 16, 1837, Harris Papers [MSciB], MCZ, vol.15; TWH to George Samouelle, November 15, 1837, Harris Papers [MSciB], MCZ, vol.15.

101. TWH to John Curtis, September 6, 1838, Harris Papers, MCZ (two nonidentical versions, a copybook and a draft).

102. John Curtis to TWH, November 26, 1838, Mus Mss, MCZ.

103. Harris's New York correspondent Edwin Willcox referred to Dr. Wilckens as a "young German" who was planning to sell his cabinet and move to a farm. See Willcox to TWH, January 30, May 24, June 18, and June 25, 1838, Mus Mss, MCZ.

104. TWH to Pickering, January 17, 1838, Harris Papers [MSciB], MCZ, vol.15.

105. TWH to Hentz, February 18, 1838, Harris Papers [MSciB], MCZ, vol.15.

106. Hentz to TWH, March 11, 1838, Harris Papers [MSciB], MCZ, vol.13.

107. TWH to Pickering, January 17, 1838, Harris Papers [MSciB], MCZ, vol.15.

108. TWH to J. E. LeConte, August 5, 1838, Harris Papers, MCZ.

109. TWH to J. E. LeConte, November 7, 1839, Harris Papers, MCZ, and Mss Dept, APS (quote from version at APS).

110. J. E. LeConte to TWH, July 13, 1840, Harris Papers [MSciB], MCZ, vol.14.

111. See Tuxen, "Entomology Systematizes and Describes,"104; Tuxen refers to Moses Harris, *An Essay Wherein Are Considered the Tendons and Membranes of the Wings of Butterflies . . .* (1767). Kirby and Spence, *Introduction to Entomology*, 3:292, 620–33, 4:349–50 (quote, 3:629), discuss the structure and function, and types of veins or nervures in the wings of insects. The authors of the *Introduction* point particularly to the work of Jurine (see below); given that Jurine had not gone beyond the Hymenoptera to develop a more general system, Kirby and Spence attempt to give an overview. Specifically in regard to the Lepidoptera, they observe, "I wonder Mr. Jones's plan of ascertaining the divisions or genera and subgenera of butterflies by the neuration of their wings has never been followed up; it would I think furnish an easy clue for the extrication of the tribes of all the *Lepidoptera*. I mean as subsidiary to more important characters." Here Kirby and Spence refer to Jones, "New Arrangement of Papilios." In his classic work, John Comstock observed that "Although the wing-characteristics have been used in the classification of insects from a very early period," they came into prominence especially in the post-Darwinian period when the subject was approached from the perspective of development from a "single primitive type;" Comstock, *Wings of Insects*, 1.

112. TWH to Hentz, February 6, 1826, Harris Papers [MSciB], MCZ, vol.13.

113. Hentz to TWH, February 19, 1826, Harris Papers [MSciB], MCZ, vol.13. Louis Jurine, *Nouvelle Methode de Classer les Hymenoptères et les Diptères* (1807) as noted by Tuxen, "Entomology Systematizes and Describes,"104.

114. TWH to Say, March 19, 1829, Harris Papers [MSciB], MCZ, vol.14 (undated), and Mss Dept, APS. In the draft (MCZ version), Harris states that "If you have any difficulty in procuring those of the last order [Lepidoptera]," Say is welcome to those that he has made. The letter sent (APS version) states the proposition: "Unless better supplied you are welcome to my drawings," which suggests that drawings of the nervures may be available elsewhere (although Harris may have meant that Say could arrange on his own to have them done).

115. TWH to Doubleday, September 15, 1839 (continued September 22, October 5, November 19), Harris Papers [MSciB], MCZ, vol.13. The Bombyces and Geometrae are groups of moths.

116. Knight, *Ordering the World,* 23–24.

117. William Sharp MacLeay presented his ideas in his *Horae Entomologicae* (London, 1819–21) and they were further developed by colleagues in the Zoological Club that emerged from the Linnaean Society in London. For this history and explanation of the system, see Winsor, *Starfish, Jellyfish, and the Order of Life,* 82–85. Knight, *Ordering the World,* 95, observed that by 1840 the quinary system was in disfavor and in fact was considered an example of the direction in which classification should not proceed.

118. TWH to Hentz, April 9, 1827, Harris Papers [MSciB], MCZ, vol.13. Harris's letter to Hentz (see transcription in T. W. Harris, 1869, *Entomological Correspondence,* 27) includes an unattributed quotation on the subject that may have been taken from Kirby and Spence.

119. TWH to J. L. LeConte, November 29, 1852, LeConte Correspondence, ANSP.

120. TWH to Hentz, March 10, 1827, Harris Papers [MSciB], MCZ, vol.13. This quotation is crossed out in Harris's draft of the letter. Harris had reported to Say the previous year that his use of the nervures to determine genera of butterflies encouraged him to extend his investigations to the Nocturna (moths). TWH to Say, May 1826, Harris Letters to Say, Houghton.

121. TWH to Hentz, June 17, 1828, Harris Papers [MSciB], MCZ, vol.13.

122. Adult insect.

123. TWH to Hentz, September 4, 1828, Harris Papers [MSciB], MCZ, vol.13. Kirby and Spence, *Introduction to Entomology,* 4:451–52, refer to the era dominated by Jan Swammerdam (1637–80) and John Ray (1627–1705) as having elevated metamorphosis as a basis for a natural classification of insects, arguing that, while metamorphosis used alone "will . . . lead to an artificial arrangement, it furnishes a very useful clue when the consideration of insects in their perfect state is added to it." Kirby and Spence likewise attempted to explain the elusive meaning of "habit" in charting animal relations. Naturalists' use of the term encompassed insect "proportions, general aspect, and figure," but they admitted that these qualities (and their use in comparing one insect to another) "though easily perceived by a practised eye, is described with much difficulty" (4:564). It appears from their definition that, for the entomologist, habit was not predominantly a matter of insect behavior.

124. TWH to Hentz, December 19, 1828, Harris Papers [MSciB], MCZ, vol.13.

125. TWH to Pickering, January 17, 1838, Harris Papers [MSciB], MCZ, vol.15; Am Acad Arts Scis Records, vol. 2 (1821–57), 115 (January 31, 1838 meeting); draft of appeal for publication support from the American Academy (undated), Harris Papers [MSciB], MCZ, vol. 2.

126. TWH to Hentz, February 18, 1838, Harris Papers [MSciB], MCZ, vol.15.

127. TWH to J. G. Morris, December [23 or 25], 1840, Harris Papers, MCZ. Harris argued for the importance of knowledge of the character of the various stages of an insect's life for all classificatory work, but considered it a requirement for the Lepidoptera. TWH to Melsheimer, January 6, 1841, Harris Papers [MSciB], MCZ, vol.12, and Harris Papers, MCZ. As a practical aid, he did not put entire reliance on the larvae as preserved in spirit but usually prepared drawings (as well as written descriptions) for future reference. TWH to Melsheimer, January 4, 1842, Harris Papers [MSciB], MCZ, vol.12, and Harris Papers, MCZ.

128. TWH to Michael Moore, Jr., October 15, 1838, Harris Papers, MCZ.

129. Ibid., where Harris says that Doubleday has "been with me some days;" TWH to Gould, October 23, 1838, Gould Papers, Houghton; and Doubleday to

TWH, November 17, 1838, Harris Papers [MSciB], MCZ, vol.13, by which date Doubleday was in New York.

130. Harris crossed out work on the Lepidoptera and replaced it with "Entomological labors."

131. TWH to Charles J. Ward (addressed to Roscoe, OH), November 8, 1838, Harris Papers [MSciB], MCZ, vol.15.

132. TWH to Melsheimer, February 24, 1839, Harris Papers, MCZ.

133. TWH to Herrick, January 15, 1839, Harris Papers [MSciB], MCZ, vol.12, and Harris Papers, MCZ.

134. TWH to Herrick, February 14, 1839, Harris Papers [MSciB], MCZ, vol.12.

135. Weiss, *Pioneer Century of American Entomology*, 107. The latter paper was T. W. Harris, 1854, "Description of Rhinosia pometella." Accounts of Harris's entomological contributions do not often relate to the number of insects that he first described and named, tending to emphasize his work on the habits and economic effects of the insects. Essig (whose interests related to California) noted that "In addition to his economic studies Harris did some excellent systematic work and described a number of important economic insects which occur throughout much of the country." He lists more than a dozen insects named by Harris that are of economic interest, especially Lepidoptera species. Essig, *History of Entomology*, 653. R. P. Dow, in his account of Harris's life, notes that he is credited with twenty-four beetle species and one genera (he refers to "the checklist," probably one by Samuel Henshaw). Dow, "Work and Times of Dr. Harris," 116.

136. T. W. Harris, 1839, "Descriptive Catalogue of the North American Insects Belonging to the Linnaean Genus Sphinx," 289.

137. TWH to Melsheimer, February 24, 1839, Harris Papers, MCZ.

138. One of Harris's genera, Ceratomia, is recognized in Holland, *Moth Book*, 47. Neither of Harris's other two genera appear there, and none of his family names appear to have been preserved.

139. See Elliott, *Biographical Dictionary*.

140. TWH to J. G. Morris, September 17, 1839, Harris Papers, MCZ. It was noted by the subsequent generation of entomologists that Harris's paper was "the first review of the American species of the Linnaean genus Sphinx in its widest sense" and that it recognized "the families now generally adopted." However, Harris's lack of access to foreign resources, especially to the works of Jacob Hubner, were noted as a deficiency. Nonetheless, "several species and quite a number of larvae are described [by Harris] . . . for the first time." In the summary outline of the group, Harris is credited with two genera and three species. Of seven Harris synonyms, four of the insects are assigned to Hubner. John B. Smith, "Monograph on the Sphingidae of America," 66–68, 238–40.

141. TWH to J. E. LeConte, November 7, 1839, Harris Papers, MCZ, and Mss Dept, APS.

142. Doubleday to TWH, June 4, 1839, Harris Papers [MSciB], MCZ, vol.13.

143. TWH to Melsheimer, November 22, 1839, Harris Papers [MSciB], MCZ, vol.12, and Harris Papers, MCZ.

144. TWH to Melsheimer, September 16, 1839, Harris Papers [MSciB], MCZ, vol.12.

145. TWH to Melsheimer, November 22, 1839, Harris Papers [MSciB], MCZ, vol.12, and Harris Papers, MCZ.

146. Harris inserts an asterisk at this point, with the following footnote: "<u>affinity</u> is relationship by marriage, which can exist only between <u>individuals</u> of a species."

147. TWH to Doubleday, September 15, 1839 (continued September 22, October 5, November 19), Harris Papers [MSciB], MCZ, vol.13.

148. Winsor, *Starfish, Jellyfish, and the Order of Life,* 82–85.

149. TWH to Doubleday, October 8, 1840, Harris Papers [MSciB], MCZ, vol.13.

150. TWH to J. G. Morris, December [23 or 25], 1840, Harris Papers, MCZ.

151. Doubleday to TWH, November 16, 1840, Harris Papers [MSciB], MCZ, vol.13.

152. TWH to J. G. Morris, December [23 or 25], 1840, Harris Papers, MCZ.

153. TWH to Doubleday, September 27, 1840, Harris Papers [MSciB], MCZ, vol.13. T. W. Harris, 1841, "Remarks on Some North American Lepidoptera."

154. TWH to Melsheimer, January 4, 1842, Harris Papers [MSciB], MCZ, vol.12, and Harris Papers, MCZ.

155. Melsheimer to TWH, February 14, 1842, Harris Papers [MSciB], MCZ, vol.12.

156. TWH to Doubleday, April 13, 1840, Harris Papers [MSciB], MCZ, vol.13.

157. TWH to Melsheimer, December 4, 1840, Harris Papers [MSciB], MCZ, vol.12, and Harris Papers, MCZ.

158. TWH to [J. E. LeConte?], November 16, 1842, Mss Dept, APS.

159. TWH to Affleck, November 26, 1846 (date replaced in pencil by December 12), Harris Papers [MSciB], MCZ, vol.14, and TWH to Affleck, December 12, 1846, Affleck Papers, LSU, which is in part derived from this draft. The quote is taken from the MCZ version (with some minor corrections of my transcription from reading the LSU version).

160. DeKay (Lyceum of Natural History, NY) to TWH, January 23, 1838, Mus Mss, MCZ.

161. TWH to DeKay, February 3, 1838, Harris Papers, MCZ.

162. DeKay to TWH, February 21, 1838, Mus Mss, MCZ.

163. TWH to William LeBaron, March 13, 1838, Harris Papers, MCZ.

164. TWH to Willcox, May 29, [1838?], Harris Papers, MCZ.

165. Willcox to TWH, April 2, 1838, Mus Mss, MCZ.

166. TWH to Willcox, April 6, 1838, Harris Papers, MCZ.

167. DeKay to TWH, June 15, 1838, Mus Mss, MCZ. Harris, in writing to Edwin Willcox on April 6, 1838, asked that an enclosed manuscript sample be transmitted to DeKay—TWH to Willcox, April 6, 1838, Harris Papers, MCZ—and Willcox later reported the favor done. Willcox to TWH, May 3, 1838, Mus Mss, MCZ.

168. DeKay to TWH, January 2, 1839, Mus Mss, MCZ.

169. TWH to J. G. Morris, December [23 or 25], 1840, Harris Papers, MCZ.

170. Meisel, *Bibliography of American Natural History,* 2:613–14.

171. Barnes, *Asa Fitch,* 39, 45.

172. For a summary account of the Massachusetts surveys in the 1830s, see Merrill, *Contributions to a History of American State Geological and Natural History Surveys,* 149–58. Merrill includes extracts from state published documents as part of this account. Also see Stone, "Role of the Learned Societies," 374–88; see 379–88, for the second survey and the role of the Boston Society of Natural History in its origins and work.

173. Meisel, *Bibliography of American Natural History,* 2:647, 2:461 (reproducing Gould's account).

174. Edward Everett to George B. Emerson, June 10, 1837, Everett Papers, MHS.

175. *Reports of the Commissioners on the Zoological Survey,* 53–54; Meisel, *Bibliography of American Natural History,* 2:647.

176. TWH to Hentz, February 18, 1838, Harris Papers [MSciB], MCZ, vol.15.

177. "Author's Preface" to second edition, in T. W. Harris, 1862, *Treatise on Some of the Insects,* v–vi.

178. Bidwell, *Rural Economy in New England*, 319, estimated that 90% of the population earned their living from farming.

179. Gates, *Farmer's Age*, 22; Thornton, *Cultivating Gentlemen*, 194.

180. Gates, *Farmer's Age*, 27–28.

181. Russell, *Long, Deep Furrow*, 257.

182. Bidwell, *Rural Economy in New England*, 345–46, 352–53; Gates, *Farmer's Age*, 256.

183. Bidwell, *Rural Economy in New England*, 322, 326, 336 (quote).

184. Bidwell and Falconer, *History of Agriculture in the Northern United States*, 260.

185. Russell, *Long, Deep Furrow*, 374–76.

186. Ibid., 292.

187. Gates, *Farmer's Age*, 269–70.

188. T. W. Harris, 1842, *Treatise on Some of the Insects*, index. Entries in the index refer to plant and vegetable organisms that are injured by insects as well as to common insect names that include the name of a vegetable (e.g., cabbage butterfly, cucumber skippers, grape-vine leaf-hopper). The index refers to each individual page on which an organism is discussed (i.e., no inclusive page references are made) and each page is counted here. The category of trees, shrubs, vines (non-fruit) includes some references to nut trees and to hop vines. Also see "Examples of Good Farming," *New England Farmer*, for examples of crop yields and dollar values from several farms in the 1820s, which gives a fair idea of the produce (including farm animals) that characterized well-run farms (some of which were recognized with a premium from the Massachusetts Society for Promoting Agriculture). The vegetables are essentially as given in Harris's report; beets and mangel-wurtzel (used for cattle feed) are among the crops that are missing from the report's index of insect damage.

189. Bidwell, *Rural Economy in New England*, 319.

190. Thornton, *Cultivating Gentlemen*, 58, 119, 180; Bidwell, *Rural Economy in New England*, 344–45.

191. Bidwell and Falconer, *History of Agriculture in the Northern United States*, 259.

192. See *National Cyclopaedia of American Biography*, 11:181–82.

193. Russell, *Agricultural Progress in Massachusetts*, 6–7. For a general discussion of the relations of agriculture, learned societies, and science in the antebellum period, see Rossiter, "Organization of Agricultural Improvement in the United States, 1785–1865." Rossiter points to the years after 1840 as a period of significant growth of interest in the improvement in agricultural practice, 291–93.

194. Russell, *Agricultural Progress in Massachusetts*, 7–8.

195. Gates, *Farmer's Age*, 343. For a study of the readership of agricultural journals, based on local practice and subscribers to one journal, see McMurry, "Who Read the Agricultural Journals?" McMurry essentially confirms the point made here but adds detail and nuance to our knowledge of journal readership.

196. Lemmer, "Early Agricultural Editors," 7–8.

197. See Russell, *Long, Deep Furrow*, index (Pests: insects), for references to historic episodes of insect damage in New England.

198. "Author's Preface" to second edition, in T. W. Harris, 1862, *Treatise on Some of the Insects*, v–vi.

199. Though there was not much time, explaining to Hentz that he intended to write up the report over the coming winter. TWH to Hentz, August 13, 1837, Harris Papers [MSciB], MCZ, vol.15. There is no evidence that Hentz ever took up the work on spiders for the Massachusetts survey and they are not included in Harris's report. In a letter to Hentz in February 1838, Harris seems to indicate he had never

heard from him on the proposition. TWH to Hentz, February 18, 1838, Harris Papers [MSciB], MCZ, vol.15.

200. TWH to Hentz, August 13, 1837, Harris Papers [MSciB], MCZ, vol.15.

201. George B. Emerson to TWH, April [21?], 1838, Harris Papers [MSciB], MCZ, vol.15; T. W. Harris, 1838, "Report on the Habits of Some Insects Injurious to Vegetation."

202. Meisel, *Bibliography of American Natural History*, 2:648.

203. TWH to Prof. Ernst F. Germar, June 25, 1838, Harris Papers [MSciB], MCZ, vol.15.

204. Massachusetts, *Resolves of the General Court*.

205. TWH to Herrick, January 20, 1840, Harris Papers [MSciB], MCZ, vol.12.

206. [Thaddeus Mason Harris] to TWH, February 8, 1840, Harris Papers [MSciB], MCZ, vol.14.

207. Massachusetts, General Court—House, House Document no. 22 (1840).

208. Merrill, *Contributions to a History of American State Geological and Natural History Surveys*, 155.

209. TWH to Doubleday, April 12, 1841, Harris Papers [MSciB], MCZ, vol.13.

210. D. H. Storer and W. O. B. Peabody published reports on Massachusetts fishes and reptiles, and on birds, respectively, in the *Journal* during 1839 and 1840.

211. TWH to George B. Emerson, February 1, 1841, Harris Papers [MSciB], MCZ, vol.14.

212. TWH to Herrick, November 24, 1841, Harris Papers [MSciB], MCZ, vol.12.

213. TWH to Herrick, January 1, 1841 [probably 1842], Harris Papers [MSciB], MCZ, vol.12.

214. TWH to Herrick, November 3, 1842, Harris Papers [MSciB], MCZ, vol.12; TWH to editor of Genesee Farmer, March 8, 1841, Harris Papers [MSciB], MCZ, vol.14 (on the lack of access to cultivation of wheat); T. W. Harris, 1842, *Treatise on Some of the Insects*, 423–24.

215. Massachusetts [Legislature], Resolve, [April 12?], 1850, Harris Papers [MSciB], MCZ, vol.14.

216. T. W. Harris, 1852, *Treatise on Some of the Insects*.

217. T. W. Harris, 1862, *Treatise on Some of the Insects*. Quote from Editor's Preface, iii.

218. OCLC's WorldCat online catalogue lists issuances for 1862, 1880, 1884, and 1890 by Orange Judd and Company. The 1841 *Report* was reprinted by Arno (New York) in 1970 in its American Environmental Studies series.

219. Discussion here is taken from the reprinting of Harris's 1841 state *Report* as T. W. Harris, 1842, *Treatise on Some of the Insects*.

220. T. W. Harris, 1842, *Treatise on Some of the Insects*, 3–4.

221. Ibid., 50.

222. Ibid., 114.

223. TWH to J. G. Morris, December [23 or 25], 1840, Harris Papers, MCZ.

224. T. W. Harris, 1842, *Treatise on Some of the Insects*, 206.

225. Ibid., 242, 243.

226. The name is attributed to Fabricius in T. W. Harris, 1869, *Entomological Correspondence*, 287.

227. T. W. Harris, 1842, *Treatise on Some of the Insects*, 247–48.

228. See *North American Review* 54 (January 1842): 73–101; *American Journal of Science and Arts* 43 (October 1842): 386–88.; *Boston Medical and Surgical Journal* 26 (April 27, 1842): 191–92.

229. T. W. Harris, 1842, *Treatise on Some of the Insects*, 152–53.

230. *Magazine of Horticulture* 9 (1843): 216–31, referencing 216–17.

231. Josiah Tatum (Philadelphia) to [TWH], June 30, 1847, with notes on soy beans, Harris Papers, HU Bot Libs.

232. Higginson, "Memoir of Thaddeus William Harris," xxix.

233. Ordish, *John Curtis*, 73 (quote), 78.

234. TWH to John Curtis, September 6, 1838, Harris Papers, MCZ (two nonidentical versions, a copybook and a draft).

235. Doubleday to TWH, April 16, 1846, Harris Papers [MSciB], MCZ, vol.13.

236. Ordish, *John Curtis*, 78–79.

237. Ordish, *Constant Pest*, 131.

238. Fitch to TWH, November 26, 1851, Harris Papers [MSciB], MCZ, vol.14. Curtis devoted the first five chapters of his *Farm Insects* to turnips. Ordish, *John Curtis*, 81.

239. Fitch, *First and Second Report on the Noxious, Beneficial and Other Insects*, 1.

240. Howard, *History of Applied Entomology*, 30–33.

241. *Family Story*, 47, 54 (quote).

242. E. D. Harris, "Memoir of Thaddeus William Harris," 319.

243. Hentz to TWH, March 1829 (and continued March 8), Harris Papers [MSciB], MCZ, vol.13.

244. Lindroth, "Systematics Specializes," 121–22.

245. TWH to Melsheimer, January 6, 1841, Harris Papers [MSciB], MCZ, vol.12, and Harris Papers, MCZ.

246. Doubleday to TWH, May 16, 1841, Harris Papers [MSciB], MCZ, vol.13.

247. Knight, *Ordering the World*, 126; *Dictionary of National Biography*, vol. 14; Rehbock, *Philosophical Naturalists*, 29. I am grateful to Fred Burchsted for providing the last reference.

248. It reads: "To / Thaddeus William Harris, M.D., / Librarian to Harvard University, / I Respectfully Dedicate / This Humble Attempt to Portray / The Principles / Pervading / The System of Nature." I am indebted to Annette E. Springer of the American Museum of Natural History Library, New York, for providing a photocopy of this dedicatory page.

249. TWH to Edward Newman, January 7, 1844, Harris Papers [MSciB], MCZ, vol.14. Higginson notes this letter as an exception to Harris's entomological correspondence during his later years, which otherwise "bore chiefly on the topics covered by [his] . . . treatise." Higginson, "Memoir of Thaddeus William Harris," xxx.

250. T. W. Harris, 1842, *Treatise on Some of the Insects*, 10–19 (quote, 10).

251. TWH to J. L. LeConte, November 25, 1851, Mss Dept, APS.

252. TWH to J. L. LeConte, December 6, 1851, LeConte Correspondence, ANSP.

253. TWH to J. L. LeConte, November 29, 1852, LeConte Correspondence, ANSP. Harris expressed some pleasure that through conversations they had, Agassiz was "coming to my opinion on this subject of genera" (apparently in regard to the question of maintaining the larger groupings rather than subdividing them into new ones) and Harris gave this as a reason why the young LeConte should take some note of the question. TWH to J. L. LeConte, November 25, 1851, Mss Dept, APS.

254. TWH to [J. E. LeConte?], November 16, 1842, Mss Dept, APS.

255. LeConte indicated that he was unable to make use of the family information which is "involved in so much obscurity." J. L. LeConte to TWH, July 1, 1846, Mus Mss, MCZ.

256. J. L. LeConte to TWH, December 3, 1851, Mus Mss, MCZ.

257. J. L. LeConte to TWH, May 28, 1855, Mus Mss, MCZ.

258. See Royal Society (Great Britain), *Catalogue of Scientific Papers*.

259. This assessment of Harris's output 1846–56 is based on the titles of his publications in "List of the Writings of Thaddeus William Harris"; Hagen, "List of Papers of Dr. T.W. Harris"; and confirmed in the author's compiled bibliography.

260. J. L. LeConte to TWH, February 22, 1847, Mus Mss, MCZ.

261. The work was out of print when LeConte wrote to Harris, but Harris sent him a copy of the second edition when it was published the following year. J. L. LeConte to TWH, December 3, 1851, Mus Mss, MCZ; TWH to J. L. LeConte, December 6, 1851, LeConte Correspondence, ANSP; TWH to J. L. LeConte, November 22, 1852, LeConte Correspondence, ANSP.

262. Fitch to TWH, December 30, 1846, Mus Mss, MCZ. Fitch refers in this letter to visiting Harris eighteen months previously. Ebenezer Emmons (Albany) to TWH, August 26, 1845, Harris Papers, HU Bot Libs; and TWH to Ebenezer Emmons (draft), n.d. [written on letter from Emmons, August 26, 1845], Harris Papers, HU Bot Libs. Emmons' letter introduced Fitch to Harris.

263. Fitch to TWH, December 30, 1846, Mus Mss, MCZ.

264. TWH to Fitch, [January?] 6, 1847, Harris Papers, MCZ; TWH to Herrick, November 16, 1846, Harris Papers [MSciB], MCZ, vol.12.

265. Fitch to TWH, August 11, 1852, Mus Mss, MCZ. Harris's letter, to which Fitch is responding, has not been seen.

266. TWH to Fitch, August 14, 1852, Harris Papers, MCZ.

267. Fitch to TWH, May 27, 1853, Mus Mss, MCZ.

268. TWH to Fitch, August 14, 1852, Harris Papers, MCZ.

269. Sorensen, *Brethren of the Net*, 110, 112. Sorensen makes no reference to Harris's work in his review of the history of the Hessian fly.

270. M. H. Morris to TWH, August 31, 1843, Harris Papers [MSciB], MCZ, vol.14.

271. M. H. Morris to TWH, [September?] 12, 1847, Harris Papers [MSciB], MCZ, vol.14.

272. T. W. Harris, 1862, *Treatise on Some of the Insects*, 582–83. As early as 1841, Harris concurred with Herrick that Morris had confused the Hessian fly with some other insect; TWH to Herrick, March 31, 1841, Harris Papers [MSciB], MCZ, vol.12. Both Herrick and Fitch agreed that Morris was not describing the Hessian fly. It is probable that what she had observed was the wheat midge. Sorensen, *Brethren of the Net*, 112.

273. TWH to M. H. Morris, December 1852, Harris Papers, MCZ.

274. TWH to Westwood, [May?] [*sic*] 1854, Harris Papers, MCZ. Harris notes that he sent eight copies of his revised *Treatise* to "friends in Europe," including the Entomological Society of London.

275. TWH to J. L. LeConte, May 5, 1853, LeConte Correspondence, ANSP.

276. See account of the expedition in Lurie, *Louis Agassiz*, 148–51.

277. TWH to Doubleday, March 24, 1849, Harris Papers [MSciB], MCZ, vol.13.

278. T. W. Harris, 1850, "Description of Some Species of Lepidoptera from the Northern Shores of Lake Superior."

279. Grote, "Rise of Practical Entomology," 79.

280. Charles Wilkes to TWH, September 28, 1848, Harris Family Papers, CHS; Pickering (Boston) to TWH, October 4, 1848, Harris Family Papers, CHS.

281. Stanton, *Great United States Exploring Expedition*, 322–23.

282. TWH to Herrick, November 16, 1846, Harris Papers [MSciB], MCZ, vol.12.

283. TWH to Fitch, [January?] 6, 1847, Harris Papers, MCZ.

284. Affleck to TWH, November 24, 1846, Harris Papers [MSciB], MCZ, vol.14.

285. Affleck to TWH, January 7, 1847, Harris Papers [MSciB], MCZ, vol.14.

286. Fitch to TWH, February 12, 1847, Mus Mss, MCZ. Fitch must have been referring to Harris's section on the Bombyces in the first edition of the *Treatise,* since he never published an independent treatment of that group of Lepidoptera (see earlier discussion).

287. Fitch to TWH, May 17, 1848, Harris Papers [MSciB], MCZ, vol.14. Ebenezer Emmons' multivolume report for the Survey devoted the last volume to insects. The preface is dated July 25, 1854, and Emmons, whose name alone appears on the title page, admits that the volume is largely a compilation from the works of other people. He writes that "I have made the freest use of Dr. Harris's excellent and practical works; and have also been very much assisted by our distinguished entomologist, Dr. Asa Fitch . . ." Emmons, *Agriculture of New York,* 5:iii–iv.

288. Fitch to TWH, August 11, 1852, Mus Mss, MCZ. Harris did not heed, or was unable to follow, Fitch's advice.

289. Massachusetts [Legislature], Resolve, [April 12?], 1850, Harris Papers [MSciB], MCZ, vol.14.

290. W. B. Calhoun, Massachusetts Secretary's Office, to TWH, July 6, 1850, Harris Papers [MSciB], MCZ, vol.14.

291. TWH to Amasa Walker, Secretary of the Commonwealth of Massachusetts, n.d. [ca.1851], Harris Papers [MSciB], MCZ, vol.14; Amasa Walker, Secretary of the Commonwealth of Massachusetts, to TWH, May 20, 1851, Harris Papers [MSciB], MCZ, vol.14.

292. Amasa Walker, Secretary of the Commonwealth of Massachusetts, to TWH, January 27, 1852, Harris Papers [MSciB], MCZ, vol.14.

293. TWH to Amasa Walker, Secretary of the Commonwealth of Massachusetts, January 28, 1852, Harris Papers [MSciB], MCZ, vol.14.

294. Brinckle, *Remarks on Entomology,* 8. In 1846, Harris had noted that "our artists" were incapable of executing useful woodcuts of insects, and he was undecided what process to use for the illustrations in his contemplated second edition. He was considering engravings on stone (not, as he pointed out, lithography, but actual engravings), which were clear but much cheaper than copper engravings. TWH to Affleck, December 12, 1846, Affleck Papers, LSU.

295. Elliott, *Biographical Dictionary; Concise Dictionary of American Biography;* Sterling and others, *Biographical Dictionary of . . . Naturalists and Environmentalists.*

296. Townend Glover to TWH, December 7, 1852, Mus Mss, MCZ.

297. Harris's letter to Glover has not been seen. The second edition of Harris's *Treatise* has 1852 as the date of publication. In a letter of February 12, 1853, Harris referred to the recent publication of the new edition of the *Treatise.* TWH to [William] Darlington, February 12, 1853, Harris Papers, MCZ. Leland Howard refers to this exchange between Harris and Glover and also assumes that it was intended for some future project. Howard, *History of Applied Entomology,* 38.

298. Townend Glover to TWH, December 27, 1852, Mus Mss, MCZ.

299. Westwood (1805–93) is called the "last entomological polyhistor" by Lindroth, "Systematics Specializes," 121. His most important work was *An Introduction to the Modern Classification of Insects,* 2 vols. (London: Longman, Orme, Brown, Green, and Longmans, 1839–40). Westwood was a founder of the Entomological Society of London in 1833 and in 1861 became Hope Professor of invertebrate zoology in Oxford University. See *Dictionary of National Biography.* He published, with Harris's close friend Edward Doubleday, the three volume, *The Genera of Diurnal Lepidoptera: Comprising Their Generic Characters, a Notice of Their Habits and Transforma-*

tions, and a Catalogue of the Species of Each Genus (London: Longman, Brown, Green, and Longmans, 1846–52). A prolific author, in 1840 he assisted with the publication of Vincenz Kollar, *A Treatise on Insects Injurious to Gardeners, Foresters, & Farmers* (London: W. Smith, 1840); originally published in German, Westwood provided notes for this translation [bibliographical record].

300. TWH to Westwood, [May?] [*sic*] 1854, Harris Papers, MCZ. This letter was also cited above in discussing the Hessian fly and Harris's relationship to Asa Fitch.

301. C. M. Saxton to TWH, January 17, 1855, Harris Family Papers, CHS.

302. TWH to Jared Sparks, January 18, 1855, Sparks Papers, Houghton.

303. Dated February 28, but not seen.

304. C. M. Saxton to TWH, March 9, 1855, Harris Family Papers, CHS.

305. E. D. Harris, "Memoir of Thaddeus William Harris," 317.

306. Howard, *History of Applied Entomology*, 35.

307. TWH to John Lowell, January 25, 1827, Harris Papers, MCZ.

308. TWH to Charles Sumner, March 8, 1850, Sumner Papers, Houghton.

309. TWH to Dr. Edmund Bailey O'Callaghan (addressed to Secretary of the State, Albany, NY), March 8, 1851, Houghton, Autograph File. No doubt "Mr. Johnson" was Benjamin Pierce Johnson (1793–1869), corresponding secretary of the New York State Agricultural Society; see *Dictionary of American Biography*. Harris must have been referring to the likely publication of his paper in the society's *Transactions*.

310. Thomas Nuttall to TWH, May 20, 1851, Harris Papers, HU Bot Libs.

311. TWH to [William] Darlington, February 12, 1853, Harris Papers, MCZ.

312. E. D. Harris, "Memoir of Thaddeus William Harris," 319.

313. T. W. Harris, 1851, "Custard Squash"; T. W. Harris, 1851, "Acorn Squash"; T. W. Harris, 1852, "Pumpkins—Squashes."

314. Higginson, "Memoir of Thaddeus William Harris," xxvi; E. D. Harris, "Memoir of Thaddeus William Harris," 319. Higginson referred to the work as "an elaborate monograph."

315. E. D. Harris, "Memoir of Thaddeus William Harris," 315–16.

316. *Daily Advertiser*, Thursday January 24, 1856 (clipping), in Sibley, Collectanea Biographica Harvardiana. This probably was written by his classmate John G. Palfrey.

317. E. D. Harris, "Memoir of Thaddeus William Harris," 316.

318. TWH to Samuel A. Eliot, December 20, 1847, Harris Family Papers (box 3, folder: Harvard—land and buildings), CHS.

319. Josiah Quincy to TWH, December 29, 1846, Harris Family Papers, CHS.

320. E. D. Harris, "Memoir of Thaddeus William Harris," 316.

321. TWH to Edward Everett, March 5, 1847, Harris Family Papers (box 3, folder: College—plate), CHS. Whether these papers were received by Harris personally is not determined, but he informed President Everett that they were then in the University Library.

322. TWH to William Simonds (draft), n.d. [after February 1848] and William Simonds to TWH, August [21?], 1849, both in Harris Family Papers (box 3, folder: Harvard Genealogy), CHS.

323. TWH to John Gorham Palfrey, May 31, 1849, Palfrey Papers, Houghton.

324. E. D. Harris, "Memoir of Thaddeus William Harris," 317; Higginson, "Memoir of Thaddeus William Harris," xxxiii. This may be the manuscript notebook, of about seventy five pages, titled "A Record of the Descendants of Cap[t] Hugh Mason of Watertown, Mass. compiled from Town, Church & Family Records by One of the Family in 1844," with the Harris Family Papers, CHS.

325. E. D. Harris, "Memoir of Thaddeus William Harris," 316.

326. Ibid., 317.

327. Ibid., 321–22.

328. TWH to Doubleday, February 1844, Harris Papers [MSciB], MCZ, vol.13.

329. Higginson, "Memoir of Thaddeus William Harris," xxv; TWH to Melheimer, January 4, 1842, Harris Papers [MSciB], MCZ, vol.12, and Harris Papers, MCZ.

330. TWH to Pickering, January 17, 1838, Harris Papers [MSciB], MCZ, vol.15.

331. Weiss, *Pioneer Century of American Entomology*, 105.

332. See, for example, Hentz to TWH, March 19, 1824, Harris Papers [MSciB], MCZ, vol.13; Hentz to TWH, April 27, 1824, Harris Papers [MSciB], MCZ, vol.13; TWH to Hentz, February 18, 1838, Harris Papers [MSciB], MCZ, vol.15. For Glover, see the discussion above.

333. Librarian's Report to the Examining Committee (by John L. Sibley), July 1, 1856, HU Overseers Reports (10.6.2), HUA, vol. 2.

334. Higginson, "Memoir of Thaddeus William Harris," xxxiii.

335. Carpenter, *First 350 Years of the Harvard University Library*, 61; Librarian's Report to the President and Fellows of Harvard College, September 5, 1832, HU College Papers, HUA, vol.5.

336. Librarian's Report to the Examining Committee (by John L. Sibley), July 1, 1856, HU Overseers Reports (10.6.2), HUA, vol. 2.

337. Librarian's Report to the Examining Committee, July 15, 1853, HU Library Reports, HUA.

338. TWH to Doubleday, January 30, 1844, Harris Papers [MSciB], MCZ, vol.13.

339. Librarian's Report to the Examining Committee (by John L. Sibley), July 1, 1856, HU Overseers Reports (10.6.2), HUA, vol. 2.

340. TWH to J. G. Morris, February 5, 1845, Harris Papers, MCZ.

341. It has not been discovered whether this was the practice in the old library in Harvard Hall.

342. TWH to President Josiah Quincy, February 24, 1840, HU College Papers, HUA, vol. 10, and HU Library Letters, HUA, vol. 1.

343. TWH to Jared Sparks, July 28, 1852, Sparks Papers, Houghton.

344. Librarian's Report, July 15/16, 1835, HU Library Reports, HUA [dated July 5], and HU College Papers, HUA, vol.7 [dated July 16].

345. TWH to Doubleday, April 12, 1841, Harris Papers [MSciB], MCZ, vol.13.

346. TWH to John L. Sibley, May 15, 1841, HU Library Letters, HUA, vol. 2; Librarian's Report to the Examining Committee (by John L. Sibley), July 11, 1856, HU Overseers Reports (10.6.2), HUA , vol. 2.

347. Librarian's Report to the Examining Committee, July 11, 1842, HU Overseers Reports (10.5), HUA, vol. 6, and HU Library Reports, HUA.

348. Librarian's Report to the Examining Committee, July 11, 1849, HU Overseers Reports (10.5), HUA, vol. 8.

349. Librarian's Report to the Examining Committee, July 10, 1850, HU Library Reports, HUA.

350. Librarian's Report to the Examining Committee (by John L. Sibley), July 11, 1856, HU Overseers Reports (10.6.2), HUA, vol. 2. Statistics compiled from notes on individual reports. Harris's successor, J. L. Sibley, gives the total of 36,000 volumes. Using the figures from Harris's annual reports, 33,430 volumes and 36,699 pamphlets were added during his tenure.

351. The latter figure is from Librarian's Report to the Examining Committee, July 9, 1844, HU Library Reports, HUA; others are from compiled statistics from all reports.

352. The numbers given here are for volumes as distinguished by Harris from pamphlets.

353. Over the course of his librarianship, about 40% of the volumes and more than 90% of the pamphlets added were gifts (from statistics compiled from annual reports). Overall, gifts accounted for about 48,000 of some 70,000 items added during Harris's administration of the library.

354. *Dictionary of American Biography; American National Biography;* Librarian's Report to the Examining Committee, July 18, 1847, HU Library Reports, HUA.

355. Sibley Library Journal, HUA. I am indebted to Brian Sullivan for access to his transcript of Sibley's library journal.

356. Librarian's Report to the Examining Committee (by John L. Sibley), July 11, 1856, HU Overseers Reports (10.6.2), HUA, vol. 2.

357. Librarian's Report to the President and Fellows of Harvard College, September 13, 1838, HU Library Reports, HUA.

358. Librarian's Report to the Committee for Annual Examination, July 14, 1846, HU Overseers Reports (10.5), HUA, vol. 7.

359. Harris notes that only three bequests were available. Librarian's Report to the Examining Committee, July 10, 1850, HU Library Reports, HUA.

360. Librarian's Report to the Examining Committee, July 16, 1852, HU Library Reports, HUA. Harris here is referring to the occupation of Old Harvard Hall by the General Court in January 1764, due to a smallpox outbreak in Boston, when the building burned and the library was consumed. Morison, *Three Centuries of Harvard* 95–96. He reiterated his plea for permanent funds for administration and collection development in the library in his next annual report, but without the suggestion of public funding. Librarian's Report to the Examining Committee, July 15, 1853, HU Library Reports, HUA.

361. Morison, *Three Centuries of Harvard,* 286–89.

362. Walton, *Three-Hundredth Anniversary of the Harvard College Library,* 31.

363. Librarian's Report to the President and Fellows of Harvard College, September 1, 1834, HU Library Reports, HUA; Walton, *Three-Hundredth Anniversary of the Harvard College Library,* 31. Harris told the university treasurer in 1848 that he had begun work on a second supplement "as soon as I had seen the first supplement through the press." TWH to Samuel A. Eliot, January 1, 1848, HU Letters to the Treasurer, HUA, vol. 7 (1846–49), 222–23.

364. Higginson, "Memoir of Thaddeus William Harris," xxiv.

365. TWH to J. L. LeConte, November 10, 1852, Harris Papers, MCZ.

366. Lucy W. Say to TWH, April 13, 1835, printed in Weiss and Ziegler, *Thomas Say,* 214–16.

367. Carpenter, *First 350 Years of the Harvard University Library,* 70–71; Walton, *Three-Hundredth Anniversary of the Harvard College Library,* 31–34.

368. Librarian's Report to the President and Fellows of Harvard College, August 29, 1840, HU College Papers, HUA, vol. 10.

369. An apparent reference to Charles Folsom (1794–1872), who was librarian of the Boston Athenaeum from 1847 to 1856. *Athenaeum Centenary,* 217.

370. TWH to Samuel A. Eliot, January 4, 1848, HU College Papers, HUA, vol. 15.

371. Kenneth Carpenter, in his historical study of the Harvard Library, noted that, while the date of implementation usually is given as 1848, existing sample drawers include cards for books acquired in 1842 and 1843 and speculated that the card catalogue might actually have begun during that time period. Carpenter, *First 350 Years of the Harvard University Library,* 70. The letter to Treasurer Eliot seems to argue against that likelihood, but the question is unsettled. The catalogue by cut

and-paste was underway by April 1848, when Harris wrote to Jared Sparks to inquire whether Sparks might sell his press to the library, stating that "We want a press here now very much in preparing our new catalogue; the titles, being cut into strips, are to be pasted, and must be pressed afterwards to keep them smooth." TWH to Jared Sparks, April 27, 1848, Sparks Papers, Houghton. This does not address the question of whether they were pasted on sheets or cards.

372. Carpenter, *First 350 Years of the Harvard University Library*, 70, 72 (photograph of Harris catalogue drawer and card); Walton, *Three-Hundredth Anniversary of the Harvard College Library*, 32–34. Library historian Jim Ranz points out that the use of slips or cards for the organization of bibliographical information in American libraries was not at all uncommon in 1850. Ranz, *Printed Book Catalogue*, 52. There are samples of Harris's card catalogue preserved in the Harvard University Archives (call no. UA III 50.15.44.8). The catalogue remained in use until 1912. Walton, *Three-Hundredth Anniversary of the Harvard College Library*, 34. In 1861, Ezra Abbot, with Charles Cutter, began an author and subject card catalogue that was intended for users of the Library, employing a card that measured 5 x 2 inches. Bentinck-Smith, *Building a Great Library*, 38; Ranz, *Printed Book Catalogue*, 53.

373. The succession of the card catalogue over the printed catalogue was a lengthy process for libraries in general. See Ranz, *Printed Book Catalogue*, e.g., 53–54, 76.

CHAPTER 4. IN THE COMMUNITY OF SCIENCE

1. DeKay to TWH, January 2, 1839, Mus Mss, MCZ.
2. TWH to Michael Moore, Jr., October 15, 1838, Harris Papers, MCZ.
3. Grote, "Rise of Practical Entomology," 79.
4. Higginson, "Memoir of Thaddeus William Harris," xxxvii. Brown's historical reputation, of course, is substantial, perhaps as much for his discovery of what came to be known as Brownian movement, of importance later in studying molecular motion, as for his botanical studies. *Concise Dictionary of Scientific Biography*.
5. TWH to Hentz, January 16, 1829, Harris Papers [MSciB], MCZ, vol.13.
6. TWH to Hentz, May 16, 1825, Harris Papers [MSciB], MCZ, vol.13; TWH to Hentz, December 19, 1828, Harris Papers [MSciB], MCZ, vol.13.
7. TWH to Say, March 19, 1829, Harris Papers [MSciB], MCZ, vol.14 (undated), and Mss Dept, APS. This letter was cited in the previous chapter, where Harris offered Say his wing nervure drawings for use in a redirected work on American entomology.
8. TWH to Hentz, November 19, 1828, Harris Papers [MSciB], MCZ, vol.13.
9. Ives published "Observations on Some of the Insects Which Infest Trees and Plants, with Hints on a Method for Their Destruction," *American Gardener's Magazine* 1, no.2 (February 1835): 52–54.
10. TWH to B. Hale Ives, March 9, 1835, Harris Papers [MSciB], MCZ, vol.14.
11. B. Hale Ives to TWH, March 13, 1835, Harris Papers [MSciB], MCZ, vol.14.
12. TWH to Charles J. Ward, March 8, 1837, Harris Papers [MSciB], MCZ, vol.15.
13. Whorton, *Before Silent Spring*, 7–8.
14. "Contributions to Entomology. No. 1," *New England Farmer* 7, no. 12 (October 10, 1828), as reprinted in T. W. Harris, 1869, *Entomological Correspondence*, 337.
15. This is the implication when he referred to T. W. Harris, 1828–1829, "Contributions to Entomology" (published in the *New England Farmer*), which "would

have been continued there if I could have hoped to excite any interest in the science among those who had the power, if not the inclination, to aid it." TWH to Storer, November 2, 1836, Harris Papers [MSciB], MCZ, vol.15.

16. This refers to T. W. Harris, 1828–29, "Contributions to Entomology."

17. TWH to Pickering, October 24, 1828, Harris Papers [MSciB], MCZ, vol.12.

18. Pickering to TWH, October 30, 1828, Harris Papers [MSciB], MCZ, vol.12.

19. TWH to Pickering, July 22, 1829, Harris Papers [MSciB], MCZ, vol.12.

20. TWH to Hentz, November 19, 1828, Harris Papers [MSciB], MCZ, vol.13.

21. Reiterated in TWH to Hentz, April 3, 1830, Harris Papers [MSciB], MCZ, vol.13.

22. There are only seven letters from 1831 (four by Harris) in my resource file.

23. Weiss, *Pioneer Century of American Entomology*, 119; BSNH, Council minutes, MSciB, meetings of February 8 and March (17?), 1832.

24. Meisel, *Bibliography of American Natural History*, 2:463, 465.

25. *History of the Massachusetts Horticultural Society*, 230.

26. "List of the Writings of Thaddeus William Harris," xl.

27. Harris later noted that the *New England Farmer* had been the leading vehicle for the promotion of scientific agriculture in the region. As for his own papers published there, "I considered the work [the *N.E. Farmer*] as suitable for such contributions as Silliman's Journal would have been—with the exception only that it had not a foreign circulation; but, as I wrote for the benefit of my own countrymen, I cared not for this want." TWH to J. L. LeConte, December 21, 1852, LeConte Correspondence, ANSP. In this retrospective, among other aspects, Harris misremembers the importance he had given at the time to review of his work by his peers in natural history, as indicated above (especially in regard to their prior presentation to the Philadelphia Academy of Natural Sciences). By 1852, when the letter to LeConte was written, Harris probably was more reconciled to his role as an aid to agricultural interests.

28. The foregoing analysis is based on Harris's bibliography and not on an examination of the publications themselves.

29. TWH to Hentz, February 6, 1826, Harris Papers [MSciB], MCZ, vol.13.

30. See, e.g., TWH to J. E. LeConte, December 17, 1829, Harris Papers [MSciB], MCZ, vol.14, and Mss Dept, APS.

31. McOuat, "Species, Rules and Meaning," e.g., 479–80, 498.

32. On the latter, see, e.g., TWH to DeKay, April 16, 1838, Harris Papers, MCZ; DeKay to TWH, June 15, 1838, Mus Mss, MCZ.

33. TWH to Say, November 18, 1824, Harris Papers [MSciB], MCZ, vol.14.

34. Greene, *American Science in the Age of Jefferson*, 10–11.

35. TWH to Hentz, January 16, 1829, Harris Papers [MSciB], MCZ, vol.13. It appears that Harris had to proceed with caution, for shortly thereafter he sent some insects to Faldermann but chose not to include nondescripts. Faldermann apparently failed, on the occasion, to meet Harris's expectations for protection of his interests. TWH to F. Faldermann, February 23, 1829, Harris Letters to Say, Houghton.

36. TWH to Pickering, July 22, 1829, Harris Papers [MSciB], MCZ, vol.12.

37. TWH to J. E. LeConte, December 17, 1829, Harris Papers [MSciB], MCZ, vol.14, and Mss Dept, APS.

38. J. E. LeConte to TWH, January 12, 1830, Harris Papers [MSciB], MCZ, vol.14.

39. Pickering to TWH, March 10, 1830, Harris Papers [MSciB], MCZ, vol.12.

40. This probably refers to P. F. M. A. Dejean, *Species General de Coleoptera* . . .

(Paris, 1825–38); volumes 1–5 (1825–31) are listed as in Harris's library. "Catalogue of the Harris Library."

41. TWH to Pickering, March 25, 1830, Harris Papers [MSciB], MCZ, vol.12. About the same time, Harris wrote to John Eatton LeConte on the subject, but he explained his practice of not sending unnamed or undescribed insects abroad in terms of the impact on American students who would be forced to consult hard to find sources in all the languages of Europe, perhaps deliberately softening his criticism of European entomologists in order not to upset LeConte. Harris, nonetheless, was pleased to note that his practice was imitated by "several other American gentlemen." TWH to J. E. LeConte, n.d. [February 19, 1830], Harris Papers [MSciB], MCZ, vol.14, and TWH to J. E. LeConte, with list of insects, February 19, 1830, Mss Dept, APS (the first, draft version, is undated and differs in language and content from that at APS, but the similarities are such that it is concluded that they are the same letter).

42. TWH to Say, March 21, 1834, Harris Papers [MSciB], MCZ, vol.14.

43. Say's recent biographer makes the point emphatically that the recognition of priority of names and promotion of the prerogative of Americans to name and describe American specimens was a fundamental part of Say's orientation as a naturalist, a stance he took not only for others but in regard to his own rights as well. Say's attitude was undoubtedly a factor in Harris's own commitment to such a practice. Stroud, *Thomas Say*, 3, 52.

44. TWH to Westwood, September 25, 1835, Harris Papers, MCZ.

45. TWH to W. S. MacLeay, June 16, 1836, Harris Papers, MCZ.

46. TWH to Herrick, February 14, 1839, Harris Papers [MSciB], MCZ, vol.12.

47. TWH to Doubleday (addressed to New York City), November 21, 1838, Harris Papers [MSciB], MCZ, vol.13. That Harris was at least semi-serious about the prospect of Doubleday and Newman coming to the United States to "join in the effort" is substantiated in his repeat of that hope in a later letter. TWH to Doubleday, May 8, 1839, Harris Papers [MSciB], MCZ, vol.13.

48. Sorensen, *Brethren of the Net*, 23.

49. Haldeman to TWH, October 31, 1842, Harris Papers [MSciB], MCZ, vol.14.

50. See, e.g., reference to TWH to Westwood, November 13, 1837, Harris Papers [MSciB], MCZ, vol.15, in the previous chapter, section on "Scientific Work and Contributions: 1836–1842: Lepidopterous Insects."

51. Haldeman to TWH, October 31, 1842, Harris Papers [MSciB], MCZ, vol.14. Also quoted in Sorensen, *Brethren of the Net*, 24.

52. TWH to Melsheimer, December 21, 1842, Harris Papers [MSciB], MCZ, vol.12, and Harris Papers, MCZ (quote). On Melsheimer as compiler, see Melsheimer to TWH, November 24, 1842, Harris Papers [MSciB], MCZ, vol.12. See Sorensen, *Brethren of the Net*, 24–25, for the published output of the Society. Its catalogue did not appear until 1853, and was issued by the Smithsonian Institution.

53. Sorensen, *Brethren of the Net*, 17, 24.

54. Ibid., 23, attributes the label to R. P. Dow (1914).

55. Characterization taken from Gould, "Notice of Some Works . . . on the Nomenclature of Zoology," 11–12; Gould does not indicate that the work was incomplete as of his review. Asa Gray reviewed the part of the *Nomenclator* dealing with rules for naming; Gray, [review of Louis Agassiz, *Nomenclator Zoologicus*].

56. Agassiz to TWH, July 25, 1849, Agassiz Papers, MCZ.

57. Agassiz to TWH, July 26, 1849, Agassiz Papers, MCZ.

58. Agassiz to TWH, August [9?], 1849, Agassiz Papers, MCZ. In the absence of Harris's initiating letter, it is difficult to analyze Agassiz's particular objections. The

available correspondence does not mention another project that Agassiz was working on at this time, namely Agassiz, "Classification of Insects from Embryological Data." The paper was first presented to the meeting of the American Association for the Advancement of Science in Cambridge in August 1849, and the published version notes that it was submitted to a "commission" consisting of Harris and C. D. Meigs. Agassiz acknowledges having received insect specimens from Harris, and notes that Harris's "extensive knowledge and acquaintance with the metamorphosis of insects" (p.10) are well known and need no further comment from him, but he does not mention any special help from Harris on the project.

59. Agassiz to TWH, August [9?], 1849, Agassiz Papers, MCZ. Erichson, "Bericht." General comments on Harris's *Report* are found on p.156, with other comments or references under particular insect groups. Harris was aware of the German review in 1846, which he brought to the attention of Edward Doubleday. TWH to Doubleday, March 20, 1846 (also dated April 1), Harris Papers [MSciB], MCZ, vol.13.

60. TWH to Westwood, May 3, 1838, Harris Papers, MCZ.

61. McOuat, "Species, Rules and Meaning," 486.

62. Ibid., 504–6, 509.

63. TWH to Willcox, April 10, 1839, Harris Papers, MCZ.

64. TWH to J. G. Morris, September 17, 1839, Harris Papers, MCZ. See also Bruce, *Launching of Modern American Science,* 251. Morris had mentioned the desirability of a meeting of entomologists in an earlier letter; J. G. Morris to TWH, July 22, 1839, Mus Mss, MCZ.

65. Meisel, *Bibliography of American Natural History,* 2:680–84; Bruce, *Launching of Modern American Science,* 252. Harris was not listed among those who took part in the Association through its April 1842 meeting; see *Reports of the First, Second, and Third Meetings of the Association of American Geologists and Naturalists,* 78–79.

66. It was at the 1847 meeting that the Association voted to become the American Association for the Advancement of Science. Bruce, *Launching of Modern American Science,* 254. Harris is listed as a founding member (from 1848) of the A.A.A.S. in Kohlstedt, *Formation of the American Scientific Community,* appendix, which suggests that he also was a member of the predecessor Association of American Geologists and Naturalists.

67. "Report on Scientific Nomenclature," quote, 424. The other members of the committee were James D. Dana (who appears to have headed the group), Samuel S. Haldeman, Charles U. Shepard, Chester Dewey, and James D. Whelpley. Gould reviewed the BAAS rules in Gould, "Notice of Some Works . . . on the Nomenclature of Zoology," 1–11. See also McOuat, "Species, Rules and Meaning," 509–10, for the general influence of the BAAS rules of 1842.

68. TWH to Willcox, January 25, 1838, Harris Papers [MSciB], MCZ, vol.15.

69. [Francis Boott] to TWH, [August?] 30, 1838, Mus Mss, MCZ.

70. TWH to J. E. LeConte, August 5, 1838, Harris Papers, MCZ. The idea of borrowing an entomologist's collection could not have been all that unusual, since, at about the same time, Harris also wrote to Frederick E. Melsheimer with the proposition to borrow Melsheimer's cabinet. TWH to Melsheimer, September 7, 1838, Harris Papers, MCZ. It is possible, but not determined, that Harris wrote to Melsheimer when his proposal to LeConte proved unsuccessful.

71. TWH to George Samouelle, November 15, 1837, Harris Papers [MSciB], MCZ, vol.15.

72. Francis Boott, to whom Harris had sent his letter for transmittal to Samouelle, apparently sent it to Curtis instead. TWH to Francis Boott, November 16,

1837, Harris Papers [MSciB], MCZ, vol.15; TWH to John Curtis, September 6, 1838, Harris Papers, MCZ (two nonidentical versions, a copybook and a draft). This exchange was noted in the previous chapter in discussing Harris's efforts to procure Lepidoptera.

73. TWH to Hentz, February 18, 1838, Harris Papers [MSciB], MCZ, vol.15. This request was noted in an earlier chapter.

74. Hentz to TWH, March 11, 1838, Harris Papers [MSciB], MCZ, vol.13.

75. TWH to Willcox, May 29, [1838?], Harris Papers, MCZ.

76. TWH to Charles J. Ward, March 8, 1837, Harris Papers [MSciB], MCZ, vol.15. Harris apparently writes that "only" two or four had the requisite skills, but the number was not legible in his manuscript. Harris goes on to describe to his collector-correspondent the best pins to use and how to use them.

77. Such an analytical exercise is, of course, methodologically difficult and the outcome can be suggestive only, but it does tend to reinforce knowledge of Harris's career from other historical sources. About 150 letters (most of them authored by Harris himself) were included in this analysis, but this was not all of Harris's letters that are extant or otherwise used in this study. Those analyzed in this way were, for the most part, from the more substantive letters among Harris's entomological papers formerly at the Museum of Science, Boston, and now at the Museum of Comparative Zoology, and in the collection of Harris Papers that already were at the MCZ. Each statement in the letters that related in some way to information exchange (e.g., actual or promised transmittal of insects, requests for information from a correspondent, thanks for a favor done) was noted as to the direction of information flow (i.e., whether Harris was provider or recipient). When counting individual statements from the analyzed letters, the results for the significant time periods in Harris's life were as follows:

	Provider	Recipient
1833–1842:	43%	57%
1843–1854:	59%	41%

(To complete the record, the figures for the period before 1833 were 54% provider and 46% recipient, but no interpretive significance is seen here. The number of his correspondents was limited in this early time period, and letters exchanged with N. M. Hentz and Charles Pickering represent a large percentage.)

78. For a related discussion of document functionality and historical events, see Elliott, "Communication and Events in History"; and Elliott, "Science at Harvard University, 1846–47", esp.455–58.

79. T. W. Harris, 1835, "Insects," Hitchcock *Report;* T. W. Harris, 1839, "Descriptive Catalogue of the North American Insects Belonging to the Linnaean Genus Sphinx"; T. W. Harris, 1842, *Treatise on Some of the Insects;* T. W. Harris, 1862, *Treatise on Some of the Insects.* Harris also cites, with apparent care and precision, a number of printed sources, especially in his *Treatise,* evidence of extensive research and familiarity with the published literature.

80. TWH to Storer, November 2, 1836, Harris Papers [MSciB], MCZ, vol.15.

81. In effect, about 40% are not listed in the name index to Meisel, *Bibliography of American Natural History,* 3:525–683. In addition to references to authors of articles and books, that index also refers to individuals who are mentioned in Meisel's historical text (for example, as officers or other active participants in scientific societies) who were not natural history authors.

82. T. W. Harris, 1842, *Treatise on Some of the Insects,* 349.

83. T. W. Harris, 1862, *Treatise on Some of the Insects*, 231–32, 455. It is not known whether Harris traveled to Connecticut specifically to observe the insects in question. No correspondence with Plumb has been seen.

84. T. W. Harris, 1842, *Treatise on Some of the Insects*, 137–38; T. W. Harris, 1862, *Treatise on Some of the Insects*, 171, 504.

85. T. W. Harris, 1842, *Treatise on Some of the Insects*, 440–42; T. W. Harris, 1862, *Treatise on Some of the Insects*, 592–97.

86. Mrs. Nancy G. Gage to her cousin Rev. John L. Sibley, July 12, 1838, Harris Papers [MSciB], MCZ, vol.14.

87. T. W. Harris, 1842, *Treatise on Some of the Insects*, 447.

88. Information on Mrs. Gage is taken from a penciled note on the back of a letter, Mrs. Nancy G. Gage to TWH, September 5, 1841, Harris Papers [MSciB], MCZ, vol.14. Regrettably, by the mid-1840s, Mrs. Gage was troubled by mental illness and thereafter spent periods of her life in mental institutions in Massachusetts and New Hampshire. See Sibley Journal, HUA. I am indebted to Brian Sullivan, who transcribed Sibley's private journal and helped to make it available online at the Harvard University Archives website, for this information.

89. As identified by Meisel, *Bibliography of American Natural History*.

90. Ward has been referred to earlier, for example when Harris was seeking help following the visit by Edward Doubleday in 1838.

91. TWH to Doubleday, September 27, 1840, Harris Papers [MSciB], MCZ, vol.13. Willcox's letters to Harris in the two years of their correspondence (1838–40) reveal an entomologically knowledgeable and personable individual who must have been a valuable source of information for Harris on the state of insect study in New York City, where Willcox was in business. In particular, he reported to Harris on the health and status of John Eatton LeConte, and never missed an opportunity to point out LeConte's weaknesses as an entomologist while telling Harris that "your reputation as an entomologist stands far above any gentleman in this country." Willcox to TWH, January 18, 1838, Mus Mss, MCZ. Though Willcox was "just of age" in early 1838, he had a number of entomological contacts, and, perhaps as an extension of his commercial activities, assisted Harris with the shipment and receipt of packages. Willcox wrote to Harris on September 25, 1838 to introduce Edward Doubleday, who was about to visit Boston. At various points, misunderstandings developed between Harris and Willcox over the terms of their relationship, in particular to Harris's claim to the insect specimens that he named for Willcox. There also are repeated references by Willcox to his desire for Oregon insects, and, as noted, his procurement of the Townsend collection was the event that finally terminated their relationship. For letters from Willcox to Harris, January 5, 1838 to February 10, 1840 (including Willcox to Harris, September 25, 1838, referred to above), see Willcox Papers, Mus Mss, MCZ.

92. Dr. John M. B. Harden to TWH, December 12, 1844, Mus Mss, MCZ.

93. This exchange was discussed in chapter 2.

94. Stephen Calverley to TWH, November 13, 1848, Mus Mss, MCZ. Calverley was an old Englishman employed as a weightmaster with a shipping company. Sorensen, *Brethren of the Net*, 166

95. Stephen Calverley to TWH, December 4, 1848, Mus Mss, MCZ. Harris's response to Calverley's letter of November 13, 1848 is not independently known but is based on what Calverley says in this subsequent letter.

96. John W. Proctor to TWH, July [7?], 1851, Harris Papers [MSciB], MCZ, vol.14.

97. TWH to John W. Proctor, September 11, 1851, Harris Papers, MCZ.

98. The apparent sequence of the events was as follows: Proctor published Harris's July 7 letter to him, in *The Salem Observer* 29, no. 28 (July 12, 1851), without Harris's prior knowledge; in the letter, Harris stated his strong views that insects were not responsible for the potato rot. It appears that Mr. Whipple published a response to this letter, and Harris again wrote to Proctor on August 6, 1851, responding to Whipple, and Proctor published it in *The Salem Observer*, 29, no. 33 (August 16, 1851), 2. Whether Whipple published a response to Harris's second letter is not determined. However, Harris's letter of September 11 to Proctor (quoted above) alludes to at least a manuscript response from Whipple. The incident is given here in the context of Harris's entomological relations to the community, but the cause of the potato rot was a matter of considerable discussion and uncertainty in the mid-nineteenth century. Probably for this reason Harris's two letters were reprinted several other times in the agricultural press. For an interesting overview of the subject, see Wheeler, "Tarnished Plant Bug." As an indication of the stakes in the potato rot discussion, Wheeler notes that Massachusetts in 1851 offered $10,000 for a cure (p.319), although this matter never seems to have emerged in Harris's involvement in the question. Wheeler refers to Harris's part in the subject, through Proctor, who is identified as president of the Essex Agricultural Society, and quotes Harris's views on the subject, concluding that "With one of America's most respected entomologists virtually dismissing insects as the cause of the rot, the insect theories gained few adherents," (p.326) although Wheeler goes on to discuss one who did pursue the matter of insect causation.

99. E.g., T. W. Harris, 1835, "Insects," Hitchcock *Report*, 601.

100. Hentz to TWH, January 1, 1826, Harris Papers [MSciB], MCZ, vol.13, quoted in Higginson, "Memoir of Thaddeus William Harris," xx.

101. TWH to Hentz, June 20, 1826 (continued on July 24, 1826), Harris Papers [MSciB], MCZ, vol.13.

102. TWH to Hentz, December 19, 1828, Harris Papers [MSciB], MCZ, vol.13.

103. Hentz to TWH, February 8, 1830 (continued April [March?] 24), Harris Papers [MSciB], MCZ, vol.13.

104. Seventy-four letters exchanged between Harris and Hentz (1824–45) have been seen, of which two-thirds are dated 1824–30.

105. TWH to Prof. Ernst F. Germar, June 25, 1838, Harris Papers [MSciB], MCZ, vol.15.

106. Hentz to TWH, April 28, 1845, Harris Papers [MSciB], MCZ, vol.13; TWH to Affleck, Dec. 12, 1846, Affleck Papers, LSU. This most likely was Hentz's "Descriptions and Figures of the Araneides of the United States," published serially in the *Boston Journal of Natural History*, beginning in 1841 and continuing to 1850. The work was reprinted in *The Spiders of the United States: A Collection of the Arachnological Writings of . . . Hentz*, Boston Society of Natural History Occasional Papers no. 2 (Boston, 1875). Elliott, *Biographical Dictionary;* Meisel, *Bibliography of American Natural History.*

107. Pickering attended Harvard College and received the A.B., Class of 1823 (1849). Elliott, *Biographical Dictionary.*

108. TWH to Hentz, February 6, 1826, Harris Papers [MSciB], MCZ, vol.13.

109. TWH to Pickering, October 24, 1828, Harris Papers [MSciB], MCZ, vol.12.

110. See, e.g., Pickering to TWH, November 11, [1826], Harris Papers [MSciB], MCZ, vol.12; TWH to Pickering, October 24, 1828, Harris Papers [MSciB], MCZ, vol.12; Pickering to TWH, October 30, 1828, Harris Papers [MSciB], MCZ, vol.12.

111. TWH to Storer, November 2, 1836, Harris Papers [MSciB], MCZ, vol.15.

112. Following his return from the Wilkes Expedition, Pickering settled in Boston.

113. Substituted for "liberal."

114. TWH to Pickering, October 6, 1836, Harris Papers [MSciB], MCZ, vol.15. In this letter, Harris also offers his opinion about a possible comparative anatomist who might accompany the Expedition.

115. TWH to Pickering, October 1836, Harris Papers [MSciB], MCZ, vol.15.

116. TWH to Pickering, December 13, 1836, Harris Papers [MSciB], MCZ, vol.15.

117. TWH to Dr. H. B. Hornbeck, August 17, 1838, Harris Papers, MCZ.

118. TWH to Doubleday, May 8, 1839, Harris Papers [MSciB], MCZ, vol.13. It was during this general time period that Harris was involved in an awkward controversy with Randall regarding ownership of the Townsend insects, discussed in the previous chapter.

119. TWH to Dr. H. B. Hornbeck, August 17, 1838, Harris Papers, MCZ.

120. William LeBaron to TWH, March 8, 1838, Mus Mss, MCZ; William LeBaron to TWH, April 21, 1841, Mus Mss, MCZ (quote); Sorensen, *Brethren of the Net*, 61–62, 87.

121. TWH to Pickering, March 3, 1828, Harris Papers [MSciB], MCZ, vol.12.

122. TWH to Leonard, February 12, 1828, Harris Papers [MSciB], MCZ, vol.14.

123. TWH to Hentz, February 26, 1828, Harris Papers [MSciB], MCZ, vol.13.

124. Fleming, *Meteorology in America*, 92, 176–77. There are no entries in Meisel, *Bibliography of American Natural History* for Levi Leonard.

125. TWH to Hentz, December 19, 1828, Harris Papers [MSciB], MCZ, vol.13.

126. TWH to Hentz, June 17, 1828, Harris Papers [MSciB], MCZ, vol.13.

127. TWH to Doubleday, May 8, 1839, Harris Papers [MSciB], MCZ, vol.13.

128. T. W. Harris, 1862, *Treatise on Some of the Insects*, 69, 171, 314–15, 540.

129. TWH to J. E. LeConte, August 5, 1838, Harris Papers, MCZ.

130. TWH to Zimmerman, November 10, 1837, Harris Papers [MSciB], MCZ, vol.15. Harris expressed the same objection directly to Schonherr, although without the reason for his wish not to have his name on American species. TWH to C. J. Schonherr, February 9, 1838, Harris Papers, MCZ.

131. John Abbot to Thaddeus Mason Harris, November 15, 1834, Mus Mss, MCZ.

132. John Abbot to TWH, November 15, 1834, Mus Mss, MCZ.

133. John Abbot to TWH, August 30, 1835, printed in *Journal of the New York Entomological Society* 22 (March, 1914): 71, and carbon of typescript in Mus Mss, MCZ.

134. TWH to Hentz, Feb. 18, 1838, Harris Papers [MSciB], MCZ, vol.15.

135. TWH to J. E. LeConte, November 7, 1839, Harris Papers, MCZ, and Mss Dept, APS. He died in late 1840 or early 1841, then in his late 80s. Elliott, *Biographical Dictionary*.

136. Elliott, *Biographical Dictionary;* Sterling and others, *Biographical Dictionary of . . . Naturalists and Environmentalists*, 1–4.

137. TWH to John Curtis, September 6, 1838, Harris Papers, MCZ (two nonidentical versions, a copybook and a draft).

138. T. W. Harris, 1839, "Descriptive Catalogue of the North American Insects Belonging to the Linnaean Genus Sphinx," 283.

139. TWH to M. H. Morris, December 16, 1854, T. W. Harris, 1869, *Entomological Correspondence*, 247–48.

140. Spencer F. Baird to TWH, February 15, 1855, Harris Family Papers, CHS.

141. Elliott, *Biographical Dictionary*.

142. TWH to Melsheimer, November 26, 1835, Harris Papers, MCZ.

143. TWH to Melsheimer, November 22, 1839, Harris Papers [MSciB], MCZ, vol.12, and Harris Papers, MCZ.

144. TWH to Doubleday, March 24, 1849, Harris Papers [MSciB], MCZ, vol.13. n the same letter, he mentioned that another good observer, Levi Leonard, no onger studied insects.

145. Referred to as Karl (Christian) Zimmermann in Sorensen, *Brethren of the Net,* ?0 and index, and refers to himself as Charles Christoph Andrew Zimmermann in Zimmermann to TWH, June 9, 1841, printed in Dow, "Work and Times of Dr. Harris," 111–13, and Weiss and Ziegler, *Thomas Say,* 207–8. In the Royal Society (Great Britain), *Catalogue of Scientific Papers.,* vols. 6, 8, and 19, he is listed as Christoph Zimmermann. For his last name, I have generally followed the form that he used when writing in English to Harris, i.e., Zimmerman.

146. A graduate of a German university, Zimmerman came to the United States n 1832 and was a corresponding member of the Entomological Society of Pennsylvania. J. L. LeConte published several of his papers after his death. Sorensen, *Brethren of the Net,* 20, 31. Harris referred to him as "Dr." TWH to Doubleday, May 8, 839, Harris Papers [MSciB], MCZ, vol.13; TWH to Dom O. J. Fahraeus, n.d., Harris Papers, MCZ.

147. *Family Story,* 51.

148. Zimmerman to TWH, May 21, 1839, Harris Papers [MSciB], MCZ, vol.14.

149. Sorensen, *Brethren of the Net,* 20.

150. TWH to Zimmerman, September 16, 1839, Harris Papers, MCZ.

151. TWH to William Oakes, October 1, 1839, Harris Papers, MCZ.

152. TWH to Melsheimer, November 22, 1839, Harris Papers [MSciB], MCZ, vol.12, and Harris Papers, MCZ.

153. TWH to Zimmerman, with letter to J. G. Morris, February 9, 1841, Harris Papers, MCZ; Zimmermann to TWH, June 9, 1841, printed in Dow, "Work and Times of Dr. Harris," 111–13, and Weiss and Ziegler, *Thomas Say,* 207–8.

154. Zimmerman to TWH, March 25, 1846, Harris Papers [MSciB], MCZ, vol.14.

155. Zimmerman to TWH, July 4, 1853, Harris Papers [MSciB], MCZ, vol.14.

156. Doubleday to TWH, April 2, 1847, Harris Papers [MSciB], MCZ, vol.13.

157. TWH to Doubleday, March 24, 1849, Harris Papers [MSciB], MCZ, vol.13.

158. For Doubleday, see *Dictionary of National Biography.*

159. T. W. Harris, 1835, "Insects," Hitchcock *Report,* 602.

160. Dix, "Notice of the Aranea aculatea," 61–63.

161. Gollaher, *Voice for the Mad,* 45, 70–71; Brown, *Dorothea Dix,* 10, 28, 48, 124 (quote); TWH to Melsheimer, December 4, 1840, Harris Papers [MSciB], MCZ, vol.12, and Harris Papers, MCZ.

162. Wade, "Friendship of Two Old-Time Naturalists."

163. Henry D. Thoreau to TWH, June 25, 1854, Harris Family Papers, CHS.

164. Harris was born in 1795 and Thoreau in 1817.

165. Wade, "Friendship of Two Old-Time Naturalists," 156.

166. Ibid., 157.

167. See Cameron, *Thoreau's Harvard Years,* part 1, p.10 and 18; also in Cameron, "Chronology of Thoreau's Harvard Years," 18. Harding, *Thoreau Handbook,* 114, includes the comment, "In natural history, T. W. Harris, the college librarian, used Smellie's *Philosophy of Natural History* and Nuttall's *Introduction to Systematic and Physiological Botany.* But both texts emphasized an aesthetic rather than a theoretical approach to nature," with a presumed outcome that would have deterred any undergraduate from a serious interest in the furtherance of natural science.

168. Canby, *Thoreau,* 50; also Canby and Raymond William Adams entry on Thoreau in the *Dictionary of American Biography,* vol.18.

169. Channing, *Thoreau the Poet-Naturalist,* 20–21.

170. Thoreau, "Natural History of Massachusetts."
171. Hildebidle, *Thoreau*, 99.
172. Van Anglen, "True Pulpit Power," 121–22 (quote, 121).
173. Thoreau, "Natural History of Massachusetts," 22.
174. Ibid., 39.
175. Van Anglen, "True Pulpit Power," 144 n. 5.
176. Dassow Walls, *Seeing New Worlds*, 128–30, 134. Dassow Walls refers to this direction in Thoreau's career as a commitment to the "Humboldtian program," 134.
177. Wade, "Friendship of Two Old-Time Naturalists," 156.
178. Thoreau, *Writings*, vol. 18: *Journal*, vol. 12:170–71; quoted in Wade, "Friendship of Two Old-Time Naturalists," 159. Wade's quotation from Thoreau journal entry, for May 1, 1859, was read against the original printed source with minor corrections.
179. Wade, "Friendship of Two Old-Time Naturalists," 152. Canby, *Thoreau*, 324 repeated the observation (though not as a direct quote) and added that Harris had stated it to Amos Bronson Alcott. Neither Wade nor Canby gives a source for the reference.
180. Dassow Walls, *Seeing New Worlds*, 6, 144–46 (quote, 145).
181. TWH to Doubleday, March 24, 1849, Harris Papers [MSciB], MCZ, vol.13.
182. Harris's draft is undated, but see Agassiz's thankful response to Harris Agassiz to TWH, [July?] 25, 1848, Harris Family Papers, CHS.
183. State 2 of Harris's letter (cited below) reads, "function of the wing-veins or nervures (neurae)."
184. TWH to Agassiz, n.d., Harris Papers [MSciB], MCZ, vol.14 (two states of the letter). Holland, *Moth Book*, 15, states: "The [wing] tubes, which are known as veins communicate with the respiratory system and are highly pneumatic. They are also connected with the circulatory system, and are furnished, at least through their basal portions, with nerves."
185. TWH to Hentz, July 28, 1829, Harris Papers [MSciB], MCZ, vol.13.
186. That Harris had not forgotten about the value of neuration is attested by William LeBaron's comment that "I agree with you in your estimate of the importance of the venation of the wings, as a means of classification." William LeBaron to TWH, October 22, 1852, Mus Mss, MCZ.
187. Agassiz to TWH, April 19, 1851, Harris Papers, HU Bot Libs.
188. TWH to Asa Gray, April 15, 1855, Harris Papers, HU Bot Libs.
189. Class of 1815 Memoranda, HUA.
190. E. D. Harris, "Memoir of Thaddeus William Harris," 322.
191. TWH to Hentz, January 19, 1827, Harris Papers [MSciB], MCZ, vol.13. In this letter, he reports his election to Hentz.
192. *History of the Massachusetts Horticultural Society*, 61, 62, 185, 210, 499–501; TWH to Hentz, March 25, 1829, Harris Papers [MSciB], MCZ, vol.13; TWH to Hentz, October 24, 1829, Harris Papers [MSciB], MCZ, vol.13.
193. TWH to Storer, June 14, 1838, Storer Papers, MSciB, pp.296–98.
194. TWH, "Entomological Lectures," Harris Papers [MSciB], MCZ, vol. 6; TWH to Gould, December 30, 1834, Gould Papers, Houghton.
195. Moses Draper to [TWH, December] 1, 1840, on the back of TWH to Marshall P. Wilder, addressed to Dorchester, December 18, 1840, Harris Papers [MSciB], MCZ, vol.14.
196. [N.?] L. Frothingham to TWH, August 4, 1842, Harris Papers [MSciB], MCZ, vol.14.

3. TWH to Doubleday, April 13, 1840, Harris Papers [MSciB], MCZ, vol. 13.

4. TWH to Doubleday, August 31, 1840, Harris Papers [MSciB], MCZ, vol. 13.

5. T. W. Harris, 1852, *Treatise on Some of the Insects,* vi.

6. TWH to Doubleday, November 17, 1842, Harris Papers [MSciB], MCZ, vol. 13.

7. Ibid.

8. FitzEdward Hall (Calcutta) to TWH, September 8, 1848, Harris Papers [MSciB], MCZ, vol. 14. The Whig Party in Massachusetts, then in control of the state legislature, likewise opposed the war. Morison and Commager, *Growth of the American Republic,* 1:612.

9. The Cotton Whigs represented the outlook of the cotton textile manufacturers who established political hegemony over commercial and merchant interests in the state during the first half of the nineteenth century. In time, the Cotton Whigs, as Harris suggested, strengthened alignment with Whigs of the South and moved away from an earlier anti-slavery stance. See Brauer, *Cotton versus Conscience,* 1–5.

10. TWH to John Gorham Palfrey, April 12, 1848, Palfrey Papers, Houghton.

11. This version is in Harris Papers [MSciB], MCZ, vol. 1, p. 275. Another version in this same source is somewhat more fanciful, noting, for example, that "Her [mother's] under dress was a blue black suit, closely fitting her plump body & covering her six limbs to the ends of her toes; and on her back she wore a kind of cloak." Also, undated manuscript draft, on back of letter from S. W. Cole (Cultivator office, Boston) to TWH, dated July 25, [1848?], Harris Papers [MSciB], MCZ, vol. 14. As Harris states, the piece was intended for publication in the Boston *Cultivator,* but a perusal of the contents of the *Cultivator* for the period from June 1848 to December 1850 has not confirmed that it was published there. In the revised edition of his *Treatise,* in discussing the great benefits derived from the lady-bird's destruction of plant lice, Harris notes that in July 1848, a friend had sent him a "whole brood of lady-bird grubs," which he raised to beetles. This undoubtedly was reference to the insects sent by Cole. See T. W. Harris, 1862, *Treatise on Some of the Insects,* 246–47. The lady-bird verse that is quoted as introductory is one of the oldest of nursery rhymes, known in varying forms.

12. E. D. Harris, "Memoir of Thaddeus William Harris;" *Family Story.*

13. Frothingham, "Memoir of Rev. Thaddeus Mason Harris," 150–151; Alan Seaburg, "Harris, Thaddeus Mason," *American National Biography.*

14. TWH (New York City) to sister, Dorothea Harris, May 20, [1822], Harris Family Correspondence, HUA.

15. TWH to Affleck, November 26, 1846 (date replaced in pencil by December 12), Harris Papers [MSciB], MCZ, vol. 14; the passage is quoted, with small modification, in TWH to Affleck, December 12, 1846, Affleck Papers, LSU.

16. First Parish Membership records, Andover-Harvard Library. The exact meaning of the category in which Harris's membership fell has not been determined. In *First Parish and First Church in Cambridge,* which lists members in 1900, the name Miss E. F. Harris, 8 Holyoke Place (pew 23) appears (p.17); this is Harris's daughter Emma Forbes Harris, who occupied the Cambridge home that Harris had built in the 1840s. *Cambridge Directory,* 1902.

17. Class of 1846 Class Book, HUA.

18. Class of 1815 Record Book, HUA.

19. TWH to Hentz, November 3, 1827, Harris Papers [MSciB], MCZ, vol. 13.

20. TWH to Doubleday, February 28, 1847, Harris Papers [MSciB], MCZ, vol. 13.

21. See, e.g., Ahlstrom, *Religious History of the American People,* 400–402.

22. *Boston Daily Courier,* Thursday January 17, 1856 (clipping), Sibley, Collectanea Biographica Harvardiana.

23. *Daily Advertiser,* Thursday January 24, 1856 (clipping), Sibley, Collectanea Biographica Harvardiana.

24. T. W. Harris, 1842, *Treatise on Some of the Insects,* 57; T. W. Harris, 1862, *Treatise on Some of the Insects,* 65.

25. See, for example, Daniels, *American Science in the Age of Jackson,* 50–53; Bozeman, *Protestants in an Age of Science,* 75; and Roberts, *Darwinism and the Divine in America,* 9. On the overall close relations of religion and nature in this era, see Novak, *Nature and Culture,* 3.

26. Bruce, *Launching of Modern American Science,* 120.

27. T. W. Harris, 1842, *Treatise on Some of the Insects,* 432–33; T. W. Harris, 1862, *Treatise on Some of the Insects,* 587. Charles Darwin came to an opposite conclusion in regard to such parasitism as evidence of benevolent design, in that he saw the suffering of the victim rather than the benefits to humans. Darwin to Asa Gray, May 22, 1860, quoted in Larson, *Summer for the Gods,* 17.

28. T. W. Harris, 1842, *Treatise on Some of the Insects,* 179–80; T.W. Harris, 1862, *Treatise on Some of the Insects,* 222.

29. Sorensen states that among agricultural entomologists, Harris was conspicuous for his use of terminology pointing to divine design in nature. He supports the point made here, however, with his observation that Harris's "occasional references seem techniques of style rather than intellectual arguments." Sorensen, *Brethren of the Net,* 64.

30. Harris Papers, MCZ [bMu 1308.41.1].

31. See, for example, TWH to B. Hale Ives, March 27, 1835, Harris Papers [MSciB], MCZ, vol. 14; TWH to Charles J. Ward, March 8, 1837, Harris Papers [MSciB], MCZ, vol. 15; TWH to Willcox, January 25, 1838, Harris Papers [MSciB], MCZ, vol. 15.

32. TWH to Melsheimer, December 4, 1840, Harris Papers [MSciB], MCZ, vol. 12, and Harris Papers, MCZ.

33. TWH to J. L. LeConte, December 6, 1846, Harris Papers, MCZ.

34. TWH to Doubleday, Sept. 15, 1839 (continued September 22, October 5, November 19), Harris Papers [MSciB], MCZ, vol. 13.

35. TWH to Hentz, February 18, 1838, Harris Papers [MSciB], MCZ, vol. 15.

36. TWH to Willcox, July 27, 1838, Harris Papers, MCZ.

37. TWH to Doubleday, September 15, 1839 (continued September 22, October 5, November 19), Harris Papers [MSciB], MCZ, vol. 13.

38. TWH to J. L. LeConte, January 23, 1849, LeConte Correspondence, ANSP.

39. TWH to J. L. LeConte, November 29, 1852, LeConte Correspondence, ANSP.

40. T. W. Harris, 1869, *Entomological Correspondence,* 267–324.

41. TWH to Melsheimer, December 4, 1840, Harris Papers [MSciB], MCZ, vol. 12, and Harris Papers, MCZ; TWH to Doubleday, August 31, 1840, T.W. Harris, 1869, *Entomological Correspondence,* 147–48.

42. Higginson, "Memoir of Thaddeus William Harris," xxv–xxvi.

43. TWH to Doubleday, August 31, 1840, T. W. Harris, 1869, *Entomological Correspondence,* 147–48.

44. TWH to Melsheimer, December 4, 1840, Harris Papers [MSciB], MCZ, vol. 12, and Harris Papers, MCZ.

45. TWH to Hon. Samuel G. Perkins, July 20, 1833, Harris Papers, MCZ.

46. Higginson, "Memoir of Thaddeus William Harris," xxvi–xxvii.

47. E. D. Harris, "Memoir of Thaddeus William Harris," 318, 321. Edward was born in 1839 and therefore was in his teens when his father died.

48. Scudder, *Butterflies,* 1:657.

49. Ibid., 1:656.

50. T. W. Harris, 1842, *Treatise on Some of the Insects,* 122; T. W. Harris, 1862, *Treatise on Some of the Insects,* 151.

51. Higginson, "Memoir of Thaddeus William Harris," xxvii.

52. TWH to Doubleday (addressed to New York City), November 21, 1838, Harris Papers [MSciB], MCZ, vol. 13.

53. E. D. Harris, "Memoir of Thaddeus William Harris," 321.

54. Novak, *Nature and Culture,* 7.

55. *Family Story,* 59.

56. TWH to Dr. William LeBaron, September 5, 1850, T. W. Harris, 1869, *Entomological Correspondence,* 260–63.

57. He considered the pinning of live specimens to be an act of cruelty, preferring to first kill them "expeditiously from humane considerations." TWH to Willcox. May 29, [1838?], Harris Papers, MCZ; TWH to Stephen Caverley, February 7, 1849, Harris Papers [MSciB], MCZ, vol. 14 (quote).

58. Sorensen, *Brethren of the Net,* chapter 5, 92–106 (see esp. 94–97), states that the idea of balance of nature, and its disruption by human action, was widely accepted in the nineteenth century and was used especially by entomologists, in relation to the questions of insect control.

59. T. W. Harris, 1862, *Treatise on Some of the Insects,* 2. In the 1842 edition of the *Treatise,* he had written only "cultivation" as the destroyer, as though at that time agriculture alone was the responsible agent resulting in the imbalance of nature. When he wrote a decade later, "civilization," a much broader term, also was responsible (as "civilization and cultivation"). In the earlier version, he noted only the disruption in the relationship between groups of insects as the result of cultivation practices, whereas in the revision for the 1852 edition (on which the posthumous edition was based) it was a much broader effect, involving relations among insects, plants, and other animals. T. W. Harris, 1842, *Treatise on Some of the Insects,* 3–4.

60. T. W. Harris, 1862, *Treatise on Some of the Insects,* 470–72 (quote, 472); T. W. Harris, 1842, *Treatise on Some of the Insects,* 340–41.

61. TWH to Asa Gray, February 28, 1850, Harris Papers, HU Bot Libs.

62. From 1852 until the late 1850s, Higginson was in Worcester, Massachusetts, as minister of the Free Church there. *American National Biography.*

63. TWH to Thomas Wentworth Higginson, "[dated a little before the death of Dr. Harris]," T. W. Harris, 1869, *Entomological Correspondence,* 263–64. The species mentioned in the last sentence are insects.

64. See, e.g., Mitchell, *Witnesses to a Vanishing America,* 14.

65. Tichi, *New World, New Earth,* viii, 2, 4.

66. Judd, *Common Lands, Common People,* 8–9, 69–72.

67. Hollerbach, "Of Sangfroid and Sphinx Moths," 201 n. 1 and 211. Hollerbach's main interest was the humane relations of nature study, and as an aside noted that Sphinx moths (with dragonflies) were the most difficult insects to kill (p.214); the former were of particular interest to Harris.

68. Nyhart, "Natural History and the 'New' Biology," 427–29. Nyhart (429–32) also traces later, post-Darwinian shifts in the relations of systematics, life histories, and morphology that are outside the parameters of this study.

69. Welch, *Book of Nature,* 5.

70. TWH to Pickering, March 25, 1830, Harris Papers [MSciB], MCZ, vol. 12.

71. TWH to B. Hale Ives, March 27, 1835, Harris Papers [MSciB], MCZ, vol. 14.

72. Welch, *Book of Nature*, 21–29, 58–60. She concentrates on Thomas Say when considering the situation in entomology and explains that problems of description and taxonomy meant that life histories were not extensively incorporated in his work.

73. Ibid., 124. This point is reiterated in Barrow, *Passion for Birds*, 74–75.

74. TWH to Zimmerman, November 10, 1837, Harris Papers [MSciB], MCZ, vol. 15 (quoted in chapter 2).

75. TWH to Doubleday, November 17, 1842, Harris Papers [MSciB], MCZ, vol. 13. See esp. Thornton, *Cultivating Gentlemen;* on Quincy and Lowell as discussed in the text, see 126–31, 135–37. Slotten, *Patronage, Practice, and the Culture of American Science*, 16–17, summarizes Democrat and Whig views in Bache's time, referring especially to Daniel Walker Howe, *The Political Culture of the American Whigs* (Chicago: University of Chicago Press, 1979).

76. Higginson, "Memoir of Thaddeus William Harris," xxix, quoting an unnamed younger entomologist.

77. E. D. Harris, "Memoir of Thaddeus William Harris," 320–21.

78. TWH to J. L. LeConte, October 13, 1851, LeConte Correspondence, ANSP.

79. Higginson, "Memoir of Thaddeus William Harris," xxvii.

CHAPTER 6. DEATH AND ASSESSMENTS

1. Librarian's Report to the Examining Committee (by John L. Sibley), July 11, 1856, HU Overseers Reports (10.6.2), HUA, vol. 2.

2. TWH to Dr. [Arminius G.] Oemler (addressed to New York City [apparently returned], November 20, 1855, Harris Papers [MSciB], MCZ, vol.14.

3. Librarian's Report to the Examining Committee (by John L. Sibley), July 11, 1856, HU Overseers Reports (10.6.2), HUA, vol. 2. T. W. Higginson described the cause of death as "pleurisy, with what is known as embolism." Higginson, "Memoir of Thaddeus William Harris," xxxiii.

4. John Langdon Sibley, Harris's assistant in the Harvard library, noted on May 30, 1846, that Harris (and his son) kept journals, where "they appear to enter very minutely into details." Harris's journal has not been located and it is not known whether he was in the habit of making entries on his personal state of mind. Sibley's own journal was largely a record and commentary on personal and public events with only occasional moments of introspection. Sibley Journal, HUA, May 30, 1846. I am grateful to Brian Sullivan, who transcribed Sibley's journal, for bringing this information to my attention.

5. As quoted below, Harris's son thought expressions of frustration and disappointment found in his father's correspondence gave an uncharacteristically negative perspective on his outlook. E. D. Harris, "Memoir of Thaddeus William Harris," 320.

6. Higginson, "Memoir of Thaddeus William Harris," xxiii, quoting from TWH to Doubleday, September 15, 1839 (continued September 22, October 5, November 19), Harris Papers [MSciB], MCZ, vol.13. Minor corrections are made here to Higginson's quote by comparing his print version to the manuscript.

7. See Hollerbach, "Of Sangfroid and Sphinx Moths," 203, 209, 211. Hollerbach's quote from Newman is from his *The Grammar of Entomology* (1835).

8. Allen, *Naturalist in Britain*, 90–93.

9. Welch, *Book of Nature,* 58.

10. T. W. Harris, 1842, *Treatise on Some of the Insects,* 19.

11. TWH to Ebenezer Emmons (draft), n.d. [written on letter from Emmons, August 26, 1845], Harris Papers, HU Bot Libs.

12. Ironically, the letter from Emmons to which Harris was responding above had as its primary function to introduce him to Fitch. See Ebenezer Emmons (Albany) to TWH, August 26, 1845, Harris Papers, HU Bot Libs.

13. Hentz to TWH, February 4, 1827, Harris Papers [MSciB], MCZ, vol.13.

14. Doubleday to TWH, April 30, 1842, Harris Papers [MSciB], MCZ, vol.13. Several years later (1846), William Spence (co-author with William Kirby of the *Introduction to Entomology*) wrote to Doubleday, who had loaned him Harris's *Treatise,* stating that he considered it "a model of what such a work ought to be." William Spence to Doubleday, [February?] 23, 1846, Harris Papers [MSciB], MCZ, vol.14 (enclosed with E. Doubleday to TWH, March 3, 1846).

15. Neave, *History of the Entomological Society of London,* 13. Westwood was, in effect, chief executive officer of the society during his tenure as secretary. Harris is listed as becoming a corresponding member in 1844, the first American other than Thomas S. Savage, who was elected a corresponding member in 1842 (160–61). (On Savage, see below.)

16. Westwood to TWH, August 10, 1845, Harris Papers [MSciB], MCZ, vol.14.

17. John Obadiah Westwood in *Dictionary of National Biography;* TWH to Westwood, September 25, 1835, Harris Papers, MCZ.

18. No doubt this was T. W. Harris, 1844, "Description of an African Beetle." The paper was read November 1, 1843, and concerned insects collected by Thomas Savage.

19. Westwood to TWH, August 10, 1845, Harris Papers [MSciB], MCZ, vol.14.

20. Thomas S. Savage to TWH, n.d. (postmarked August 4, 1847), Harris Papers [MSciB], MCZ, vol.14. Thomas S. Savage (1804–80) was an American Episcopal minister and a missionary in Liberia in the 1830s and 1840s. He published papers on African natural history, including insects, and is most noted for his descriptions of the gorilla. Elliott, *Biographical Dictionary.* Westwood's paper was "Descriptions of Two New Goliath Beetles from Cape Palmas, in the Collection of Rev. F. W. Hope," *Transactions of the Entomological Society of London* 5 (1847–49): 18–20, where he wrote of his Megalorhina Harrisii: "This magnificent insect is here named in compliment to Dr. T. W. Harris, one of the most zealous and learned Entomologists of North America" (p.20). The other insect in Westwood's paper was named for Thomas Say.

21. J. V. C. Smith to TWH, January 20, 1844, Harris Papers [MSciB], MCZ, vol.14. Jerome Van Crowninshield Smith was editor of the *Boston Medical and Surgical Journal,* where Harris's *Treatise* was positively reviewed and compared in similar terms to some of the other (unnamed) survey reports (vol. 26 [April 27, 1842]: 191–92).

22. A. Langden Elwyn to TWH, September [16?], 1848, Harris Papers [MSciB], MCZ, vol.14.

23. TWH to William Buckminster, n.d., Harris Papers [MSciB], MCZ, vol.14.

24. *Daily Advertiser,* Thursday January 24, 1856 (clipping), in Sibley, Collectanea Biographica Harvardiana.

25. E. D. Harris, "Memoir of Thaddeus William Harris," 320–21. Harris's son Edward Doubleday Harris, the author of this memoir, was born in 1839 and his father died in 1856. Insofar as he draws on his own impressions here, they are those of a child and adolescent.

26. *Family Story,* 50–51 (quote, 50).

27. Higginson, "Memoir of Thaddeus William Harris," xxxvii.

28. E. D. Harris, "Memoir of Thaddeus William Harris," 321.

29. For example, in 1838 he paid for the extra printing costs for a paper by hi young protégé John Witt Randall, because of bad Latin, even though he had op posed use of that language in the paper in the first place, but having made his poin in a private letter he wrote, "This is my excuse to you—but I do not wish to make any noise about it, or to cause any difficulty or hard thoughts on the part of any one." TWH to ?, April 30, 1838, Harris Papers, MCZ.

30. *Boston Daily Courier,* Thursday January 17, 1856 (clipping), in Sibley, Collec tanea Biographica Harvardiana.

31. *Daily Advertiser,* Thursday January 17, 1856 (clipping), in Sibley, Collectane Biographica Harvardiana.

32. *Daily Advertiser,* Thursday January 24, 1856 (clipping), in Sibley, Collectane Biographica Harvardiana. Authorship by John G. Palfrey is based on a note with the clipping.

33. *Evening Transcript,* Saturday January 19, 1856 (clipping), in Sibley, Collecta nea Biographica Harvardiana.

34. American Academy of Arts and Sciences, *Proceedings* 3 (1856): 224–25 Gould was appointed a committee to prepare a memoir of Harris's life, which re grettably he never accomplished.

35. Wade, "Friendship of Two Old-Time Naturalists," 157. Agassiz's opinion was quoted in Harris's newspaper obituary. *Daily Advertiser,* Thursday January 17 1856 (clipping), in Sibley, Collectanea Biographica Harvardiana.

36. Higginson, *Contemporaries,* 2:192; this source reproduces (with some omis sions and changes) Higginson, "Memoir of Thaddeus William Harris," where he does not give the name of the botanist.

37. Bouvé, "Historical Sketch of the Boston Society of Natural History," 64–65

38. Essex Institute, *Proceedings* 2 (1862): 2–3.

39. From the title page.

40. This is probably William Sharswood (1836–1905), who graduated from the University of Pennsylvania in 1856 and received a Ph.D. from the University of Jena in 1859. His scientific interests were especially in chemistry. It is not determined how he became interested in the republication of Harris's works. *Appleton's Cyclopae dia,* vol. 5: 483–84; Allibone, *Critical Dictionary,* vol. 2.

41. "Prospectus."

42. T. W. Harris, 1869, *Entomological Correspondence.*

43. Higginson, "Memoir of Thaddeus William Harris," xxxvii.

44. Scudder, *Butterflies,* 1:656–57.

45. Grote, "Rise of Practical Entomology," 76.

46. Howard, "Progress in Economic Entomology," 137.

47. Grote, "Rise of Practical Entomology," 77.

48. Ibid., 81.

49. Sorensen, *Brethren of the Net,* 60–61.

50. Grote, "Rise of Practical Entomology," 80.

51. From page references to the *Treatise* that Grote subsequently cites, it is es tablished that it was the first (1842) edition of the *Treatise* that Harris gave to Fitch and which is under discussion.

52. Grote states that it included "a few marginalia by both." Grote, "Rise of Practical Entomology," 80.

53. Ibid., 80–81. Here Grote apparently meant that Harris mistakes insects in

James E. Smith and Abbot, *Rarer Lepidopterous Insects of Georgia*, as the same as those in New England. Grote says that this point was first made by Alpheus Spring Packard.

54. Sterling and others, *Biographical Dictionary of . . . Naturalists and Environmentalists*, 383.

55. Howard, *History of Applied Entomology*, 34–35 (quote, 35).

56. Howard, "Progress in Economic Entomology," 136–37.

57. Howard, *History of Applied Entomology*, 30–31 (quote, 30). An argument might be made that Howard was arguing in favor of a more professionalized entomological corps for his bureau by pointing out Harris's naturalist grounding and lack of informed know-how in regard to agriculture. But Howard also makes the point that Harris's studies of insect life histories laid the basis of important work after him. In this regard, Keir Sterling notes that Howard was particularly interested in the biological control of insects, in spite of the increasing dominance of chemical means from the 1870s. Sterling and others, *Biographical Dictionary of . . . Naturalists and Environmentalists*, 383–84; Palladino, *Entomology, Ecology and Agriculture*, 29.

58. Howard, *History of Applied Entomology*, 44–45. Of the three entomologists, only Fitch appears in the *Dictionary of Scientific Biography*, attributable perhaps to the fact that at the time of preparation of the *DSB*, author Samuel Rezneck had written several items on Fitch.

59. Howard, "Progress in Economic Entomology," 137.

60. Howard, *History of Applied Entomology*, 34.

61. Quoted in Welch, *Book of Nature*, 1. Welch notes that Comstock had previously been interested in botany (p.2). Mallis, *American Entomologists*, 31.

62. Sterling and others, *Biographical Dictionary of . . . Naturalists and Environmentalists*, 383–84, 164–66.

63. Howard, "Progress in Economic Entomology," 137.

64. Whorton, *Before Silent Spring*, 14, 17–21; Sorensen, *Brethren of the Net*, 106, 124–25. See also Palladino, *Entomology, Ecology and Agriculture*, chapter 1, 21–46, esp. 21–31, for an overview of the shift to chemical remedies. Sawyer, *To Make a Spotless Orange*, xix, argues that the timing of the transition to chemical control of insects cannot be very precisely stated and seems to indicate that it was a twentieth-century phenomenon.

65. Howard, "Progress in Economic Entomology," 137–38.

66. Strecker, *Butterflies and Moths of North America*, 239.

67. Scudder, *Butterflies*, 1:656.

68. Grote, "Rise of Practical Entomology," 79.

69. "Thaddeus William Harris, M.D. [note accompanying portrait]."

70. Weiss, *Pioneer Century of American Entomology*, 108.

71. Osborn, *Fragments of Entomological History*, 310.

72. Mallis, *American Entomologists*, 31.

73. TWH to Charles J. Ward, March 8, 1837, Harris Papers [MSciB], MCZ, vol.15. It is quoted from in chapter 4.

74. "Insect Collection of Thaddeus W. Harris." See also Purrington and Nielsen, "Discovery of the T.W. Harris Collection." According to the authors of the latter article, for some time after its transfer to the Museum of Comparative Zoology, many entomologists had thought the collection was lost. They report on their rediscovery of it and their study of eleven clearwing moth species which had been described by Harris. They refer to the collection as "the oldest synoptic collection of insects in North America," and note that a number of the specimens at the time of their study were not in good condition.

75. The inscription is quoted in chapter 2.

76. American Academy of Arts and Sciences, *Proceedings* 3 (1856): 224.

77. Matthews, "History of the Cambridge Entomological Society," 15–16.

78. Mallis, *American Entomologists*, 25.

79. See, for example, Higginson's comment (quoted above) that because of his dedication and his work methods Harris's observations never came into question; also his quote from an unnamed younger entomologist, that Harris's fairness meant that he "never allowed undue weight to any set of observations, even when they were his own." Higginson, "Memoir of Thaddeus William Harris," xxxvii, xxix. Scudder (as quoted in the previous chapter) repeated the implicit belief that no one ever doubted the accuracy of Harris's work—Scudder, *Butterflies*, 1:657—, as did Howard in his statement that "Many new insect pests have been studied since his time, and many new and practical ideas in regard to the warfare against injurious insects have been advanced; but no one has had to do over the work which Harris did so well." Howard, "Progress in Economic Entomology," 137. Also see chapter 4, for the discussion of Augustus Radcliffe Grote's use of Harris—derived from familiarity with his writings—as a model for right conduct among scientists.

Bibliography

Agassiz, Louis. "The Classification of Insects from Embryological Data." *Smithsonian Contributions to Knowledge* 2 article 6 (1851).

Ahlstrom, Sydney E. *A Religious History of the American People*. New Haven: Yale University Press, 1972.

Allen, David Elliston. *The Naturalist in Britain: A Social History*. London: A. Lane, 1976; Princeton, NJ: Princeton University Press, 1994.

Allibone, S. Austin. *A Critical Dictionary of English Literature and British and American Authors*, vol. 2. Philadelphia: J. B. Lippincott, 1871.

American Academy of Arts and Sciences. *Proceedings* 3 (1856): 224–25 (resolution on death of Harris).

American Journal of Science and Arts 43 (October 1842): 386–88 (review of Harris, *Report on the Insects of Massachusetts, Injurious to Vegetation*, by "C.").

American National Biography. New York: Oxford University Press, 1999.

Appleton's Cyclopaedia of American Biography. Edited by James Grant Wilson and John Fiske. New York: D. Appleton, 1888; Detroit: Gale Research Co., 1968.

Athenaeum Centenary. The Influence and History of the Boston Athenaeum from 1807 to 1907 . . . [Boston]: The Boston Athenaeum, 1907.

Bailyn, Bernard, Donald Fleming, Oscar Handlin, and Stephan Thernstrom. *Glimpses of the Harvard Past*. Cambridge, MA: Harvard University Press, 1986.

Barnes, Jeffrey K. *Asa Fitch and the Emergence of American Entomology: with an Entomological Bibliography and a Catalog of Taxonomic Names and Type Species*. New York State Museum Bulletin no. 461. Albany: State Education Department, University of the State of New York, 1988.

Barrow, Jr., Mark V. *A Passion for Birds: American Ornithology after Audubon*. Princeton, NJ: Princeton University Press, 1998.

Barrows, Samuel J. "Dorchester in the Last Hundred Years." In *The Memorial History of Boston* . . . , edited by Justin Winsor, 3:589–600. Boston: Ticknor and Company, 1886.

Beardsley, Edward H. *The Rise of the American Chemical Profession, 1850–1900*. University of Florida Monographs, Social Sciences, no. 23. Gainesville: University of Florida Press, 1964.

Beaver, Donald deB. "The American Scientific Community, 1800–1860: A Statistical–Historical Study." Ph.D. diss., Yale University, 1966.

Bentinck-Smith, William. *Building a Great Library: The Coolidge Years at Harvard*. Cambridge, MA: Harvard University Library, 1976.

Bidwell, Percy Wells. *Rural Economy in New England at the Beginning of the Nineteenth Century*. Transactions of the Connecticut Academy of Arts and Sciences 20 (May 1916): 241–399; New Haven, CT: 1916.

Bidwell, Percy Wells and John I. Falconer. *History of Agriculture in the Northern United States 1620–1860*. Washington: Carnegie Institution of Washington, 1925; Clifton, NJ: Augustus M. Kelley Publishers, 1973.

Bigelow, Jacob. *Florula Bostoniensis: A Collection of Plants of Boston and Its Environs, with Their Generic and Specific Characters, Synonyms, Descriptions, Places of Growth, and Time of Flowering, and Occasional Remarks*. Boston: Cummings and Hilliard; Cambridge: Hilliard and Metcalf, 1814.

Boston Medical and Surgical Journal 26 (April 27, 1842): 191–92 (review of Harris, *Report on the Insects of Massachusetts, Injurious to Vegetation*).

Boston Society of Natural History. *Proceedings* 7 (1859): 72 (note on the Harris insect collection at the Society).

Bouvé, Thomas T. "Historical Sketch of the Boston Society of Natural History, with a Notice of the Linnaean Society Which Preceded It." In *Anniversary Memoirs of the Boston Society of Natural History: Published in Celebration of the Fiftieth Anniversary of the Society's Foundation, 1830–1880*. Boston: Published by the Society, 1880.

Bozeman, Theodore Dwight. *Protestants in an Age of Science: The Baconian Ideal and Antebellum American Religious Thought*. Chapel Hill: University of North Carolina Press, 1977.

Brauer, Kinley J. *Cotton versus Conscience: Massachusetts Whig Politics and Southwestern Expansion, 1843–1848*. Lexington: University of Kentucky Press, 1967.

Brinckle, William D. *Remarks on Entomology, Chiefly in Reference to an Agricultural Benefit*. Lancaster, PA: W. B. Wiley, Printer, 1852.

Brown, Thomas J. *Dorothea Dix: New England Reformer*. Cambridge, MA: Harvard University Press, 1998.

Bruce, Robert V. *The Launching of Modern American Science 1846–1876*. New York: Alfred A. Knopf, 1987.

Cambridge Directory. Boston: W. A. Greenough, 1902, 1926, 1930.

Cameron, Kenneth Walter. "Chronology of Thoreau's Harvard Years." *Emerson Society Quarterly* no. 15 (October 1959): 2–108.

———. *Thoreau's Harvard Years: Materials Introductory to New Explorations: Record of Fact and Background*. Hartford: Transcendental Books, [1966].

Canby, Henry Seidel. *Thoreau*. Boston: Houghton Mifflin Co., 1939.

Carpenter, Kenneth E. *The First 350 Years of the Harvard University Library: Description of an Exhibition*. Cambridge, MA: Harvard University Library, 1986.

"Catalogue of the Harris Library." Boston Society of Natural History. *Proceedings* 7 (1859–61): 266–71.

Channing, William Ellery. *Thoreau the Poet-Naturalist with Memorial Verses*. New ed. en. Edited by F. B. Sanborn. Boston: Charles E. Goodspeed, 1902.

Cohen, I. Bernard. *Some Early Tools of American Science: An Account of the Early Scientific Instruments and Mineralogical and Biological Collections in Harvard University*. Cambridge: Harvard University Press, 1950; New York: Russell & Russell, 1967.

Comstock, John. *The Wings of Insects*. Ithaca, NY: Comstock Publishing Co., 1918.

Concise Dictionary of American Biography. New York: Charles Scribner's Sons, 1964.

Concise Dictionary of Scientific Biography. New York: Charles Scribner's Sons, 1981.

Cravens, Hamilton. "American Science Comes of Age: An Institutional Perspective, 1850–1930." *American Studies* 17, no. 2 (Fall 1976): 49–70.

Cravens, Hamilton and Alan I. Marcus. "Introduction: Technical Knowledge in

American Culture: An Analysis," In *Technical Knowledge in American Culture: Science, Technology, and Medicine,* edited by Hamilton Cravens, Alan I. Marcus, and David M. Katzman, 1–18. Tuscaloosa: University of Alabama Press, 1996.

Daniels, George H. *American Science in the Age of Jackson.* New York and London: Columbia University Press, 1968.

———. "The Process of Professionalization in American Science: The Emergent Period, 1820–1860." *Isis* 58, no. 2 (Summer 1967): 151–66.

Dassow Walls, Laura. *Seeing New Worlds: Henry David Thoreau and Nineteenth-Century Natural Science.* Madison: University of Wisconsin Press, 1995.

DeKay, James E. *Anniversary Address on the Progress of the Natural Sciences in the United States, Delivered before the Lyceum of Natural History of New York* (New York: G. G. Carwell, 1826). Excerpted in *Science in America: Historical Selections,* edited by John C. Burnham, 75–89. New York: Holt, Rinehart and Winston, 1971.

Dictionary of American Biography. New York: Scribner's, 1928–1937.

Dictionary of National Biography. Edited by Leslie Stephen and Sidney Lee. London: Oxford University Press [1949–50].

Dix, D. L. "Notice of the Aranea aculatea, the Phalaena antiqua and some species of the Papilio." *American Journal of Science and Arts* 19 (October 1830): 61–63.

Dow, R. P. "The Work and Times of Dr. Harris." Brooklyn Entomological Society. *Bulletin* 8 (December 1913): 106–18.

Dupree, A. Hunter. *Asa Gray: 1810–1888.* Cambridge, MA: Belknap Press of Harvard University Press, 1959.

Elliott, Clark A. "The American Scientist, 1800–1863: His Origins, Career, and Interests." Ph.D. diss., Case Western Reserve University, 1970.

———. "The American Scientist in Antebellum Society: A Quantitative View." *Social Studies of Science* 5 (1975): 93–108.

———. *Biographical Dictionary of American Science: The Seventeenth through the Nineteenth Centuries.* Westport, CT: Greenwood Press, 1979.

———. "Communication and Events in History: Toward a Theory for Documenting the Past." *American Archivist* 48 (Fall 1985): 357–68 and 49 (Winter 1986): 95 (correction).

———. *The History of Science in the United States: A Chronology and Research Guide.* New York: Garland Publishing, 1996.

———. "Introduction: The Scientist in American Society." In *Biographical Dictionary of American Science: The Seventeenth through the Nineteenth Centuries,* by Clark A. Elliott, 3–9. Westport, CT: Greenwood Press, 1979.

———. "Models of the American Scientist: A Look at Collective Biography." *Isis* 73 (March 1982): 77–93.

———. "Science at Harvard University, 1846–47: A Case Study of the Character and Functions of Written Documents." *American Archivist* 57 (Summer 1994): 448–60.

Elliott, Clark A. and Margaret W. Rossiter, eds. *Science at Harvard University: Historical Perspectives.* Bethlehem: Lehigh University Press: 1992.

Emmons, Ebenezer. *Agriculture of New York.* Albany: Printed by C. Van Benthuysen & Co., 1846–54; the last volume (number 5, 1854) is devoted to insects.

Erichson, Wilhelm Ferdinand. "Bericht uber die wissenschaftlichen Leistungen in der Naturgeschichte der Insecten, Arachniden, Crustacaen u. Entomostraceen wahrend des Jahres 1842." *Archiv fur Naturgeschichte* 9, vol. 2 (1843): 149–288.

Essex Institute. *Proceedings* 2 (1862): 2–3 (death notice of Harris).

Essig, E. O. *A History of Entomology.* New York: Macmillan Company, 1931.

"Examples of Good Farming." *New England Farmer* 8, no. 32 (February 26, 1830): 249–51.

A Family Story (n.p., n.d.) (book is by Elizabeth Harris [1844–1939], apparently published ca.1908).

Farber, Paul Lawrence. "The Type Concept in Zoology during the First Half of the Nineteenth Century." *Journal of the History of Biology* 9, no. 1 (Spring 1976): 93–119.

Field, W. L. W. "The Harris Memorial Tablet." *Psyche* 17 (February 1910): 28.

First Parish and First Church in Cambridge (1636), Unitarian since 1829: List of the Officers of the Church, Parish, and Congregation and of the Members of the Congregation. Cambridge: Printed by O. B. Graves, 1900.

Fitch, Asa. *First and Second Report on the Noxious, Beneficial and Other Insects of the State of New York.* Albany: C. Van Benthuysen, 1856.

Fleming, James Rodger. *Meteorology in America, 1800–1870.* Baltimore: Johns Hopkins University Press, 1990.

Frothingham, Nathaniel L. "Memoir of Rev. Thaddeus Mason Harris,." Massachusetts Historical Society. *Collections* 4th ser., 2 (1854): 130–55.

Gates, Paul W. *The Farmer's Age: Agriculture 1815–1860.* The Economic History of the United States, vol. 3. New York: Holt, Rinehart and Winston, 1960.

Gillispie, Charles C. *The Professionalization of Science: France 1770–1830 Compared to the United States 1910–1970.* 3rd Neesima Lectures. Kyoto, Japan: Doshisha University Press, 1983.

Goldstein, Daniel. " 'Yours for science': The Smithsonian Institution's Correspondents and the Shape of Scientific Community in Nineteenth-Century America." *Isis* 85 (1994): 573–99.

Gollaher, David. *Voice for the Mad: The Life of Dorothea Dix.* New York: Free Press, 1995.

Gould, Augustus A. "Notice of Some Works, recently published, on the Nomenclature of Zoology." *American Journal of Science and Arts* 45 (October 1843): 1–12 (1–11, review of report of committee of the British Association for the Advancement of Science; 11–12, of Louis Agassiz's *Nomenclator Zoologicus*).

————. "Notice of the Origin, Progress and Present Condition of the Boston Society of Natural History," *American Quarterly Register* 14, no. 3 (1842): 236–41 (as quoted in Meisel, *Bibliography of American Natural History*, 2:457–61).

Graustein, Jeannette E. "Natural History at Harvard College, 1788–1842." Cambridge Historical Society. *Publications* 38 (1961): 69–86.

————. *Thomas Nuttall, Naturalist: Explorations in America 1808–1841.* Cambridge, MA: Harvard University Press, 1967.

Gray, Asa. [review of Louis Agassiz, *Nomenclator Zoologicus,* dealing with rules for naming]. *American Journal of Science and Arts* 2nd ser. 3 (March 1847): 302–9.

Greene, John C. *American Science in the Age of Jefferson.* Ames: Iowa State University Press, 1984.

Grote, Augustus Radcliffe. "The Rise of Practical Entomology in America." Entomological Society of Ontario. *Annual Report* 20, 1889 (1890): 75–82.

Guralnick, Stanley M. "The American Scientist in Higher Education, 1820–1910."

In *The Sciences in the American Context: New Perspectives*, edited by Nathan Reingold, 99–141. Washington, DC: Smithsonian Institution Press, 1979.

———. *Science and the Ante-Bellum American College*. American Philosophical Society Memoirs, vol. 109. Philadelphia: American Philosophical Society, 1975.

Hagen, H. A. "List of Papers of Dr. T.W. Harris, Not Mentioned in the List of His Writings in the 'Entomological Correspondence.'" Boston Society of Natural History. *Proceedings* 21 (1881): 150–52.

Hales, John G. *A Survey of Boston and Its Vicinity*. Boston: Ezra Lincoln, 1821.

Harding, Walter. *A Thoreau Handbook*. New York: New York University Press, 1959.

Hardy, Alan [editor and compiler]. "T.W. Harris and Cremastocheilus." *Scarabaeus* no. 7 (March 1983): 1–7, and no. 8 (January 1984): 1–9.

Harris, Edward D. "Memoir of Thaddeus William Harris." Massachusetts Historical Society. *Proceedings*. 19 (May 1882): 313–22 (also, tabular charts of maternal and paternal ancestry and T. W. Harris's descendants, identified as "Prepared only for Private Distribution, by Edw. Doubleday Harris, 1882," in reprint of the "Memoir" in Widener Library, Harvard University, call no. KF29122).

———., comp. *New England Ancestors of Katherine-Brattle and William-Cary Harris*. [New York: C. H. Ludwig, 1887; privately printed].

Harris, Seymour. *Economics of Harvard*. New York: McGraw-Hill, 1970.

Harris, Thaddeus Mason. *The Natural History of the Bible: A Description of All the Quadrupeds, Birds, Fishes, Reptiles, and Insects, Trees, Plants, Flowers, Gums and Precious Stones Mentioned in the Sacred Scriptures, Collected from the Best Authorities*. Boston: Wells and Lilly, 1820.

Harris, Thaddeus William. "Acorn Squash." *New England Farmer*, n.s., 3, no. 23 (November 8, 1851): 366–67.

———. "Contributions to Entomology." *New England Farmer* 7, no. 12 (October 10, 1828): 90–91; no. 15 (October 31, 1828): 117–18; no. 16 (November 7, 1828): 122–23; no. 17 (November 14, 1828): 132; no. 20 (December 5, 1828): 156; no. 21 (December 12, 1828): 164; 8, no. 1 (July 24, 1829): 1–3. Reproduced in: Harris, *Entomological Correspondence*, 337–59.

———. "Custard Squash." *New England Farmer*, n.s., 3, no. 4 (February 15, 1851): 59.

———. "Description of an African Beetle, Allied to Scarabaeus polyphemus, with Remarks upon Some Other Insects of the Same Group." *Boston Journal of Natural History* 4, no. 4 (January 1844): 397–405; extracts of the paper appeared in Boston Society of Natural History. *Proceedings* 1 (1843): 151–53.

———. "Description of a Nondescript Species of the Genus Condylura." *Boston Journal of Philosophy and the Arts* 2, no. 6 (May 1825): 580–83.

———. "Description of Rhinosia pometella." Boston Society of Natural History. *Proceedings* 4 (February 1854): 349–51.

———. "Description of Some Species of Lepidoptera from the Northern Shores of Lake Superior." In Louis Agassiz and J. Elliot Cabot. *Lake Superior: Its Physical Character, Vegetation, and Animals, Compared with That of Other and Similar Regions*, 386–94 and pl. 7. Boston: Gould, Kendall & Lincoln, 1850.

———. "Description of Three Species of the Genus Cremastocheilus." Academy of Natural Sciences of Philadelphia. *Journal* 5 pt. 2 (February 1827): 381–89.

———. "Descriptive Catalogue of the North American Insects Belonging to the

Linnaean Genus Sphinx in the Cabinet of the Author." *American Journal of Science and Arts* 36, no. 2 (1839): 282–320.

———. *A Discourse Delivered Before the Massachusetts Horticultural Society on the Celebration of Its Fourth Anniversary, October 3, 1832* [with Society proceedings and membership list]. Cambridge: E. W. Metcalf and Company, 1832, 3–54; reprinted in *New England Farmer* 11 [n.s., 2], nos. 26–32 (January 9–February 20, 1833).

———. *Entomological Correspondence of Thaddeus William Harris, M.D.* Edited by Samuel H. Scudder. Boston Society of Natural History Occasional Papers,1. Boston: Boston Society of Natural History, 1869.

———. "Insects." In Edward Hitchcock. *Report on the Geology, Mineralogy, Botany and Zoology of Massachusetts*, 566–95. Amherst: Press of J. S. & C. Adams, 1833.

———. "Insects." In Edward Hitchcock, *Report on the Geology, Mineralogy, Botany, and Zoology of Massachusetts, made and published by order of the government of that state: in four parts*, 2nd ed., cor. and enl., 553–602. Amherst: J. S. & C. Adams, 1835.

———. "List of Native Plants Discovered Growing Near Boston, the Present Season, in a letter read before the Massachusetts Horticultural Society." *Magazine of Horticulture* 6, no. 7 (July 1840): 245–47.

———. "On the History and Nomenclature of Some Cultivated Vegetables (abstract)." American Association for the Advancement of Science. *Proceedings* 5 (1851): 180–82.

———. "Pumpkins—Squashes." *New England Farmer*, n.s., 4, no. 2 (February 1852): 58–59.

———. "Remarks on Some North American Lepidoptera, by Edward Doubleday, Esq., including a communication from T. W. Harris, M.D., of Boston, U.S." *Entomologist* [London], no. 7 (May 1841): 99–101.

———. "Report on the Habits of Some Insects Injurious to Vegetation, in Massachusetts [Coleoptera]." In *Reports of the Commissioners on the Zoological Survey of the State*. Massachusetts House Document No. 72. Boston: Dutton and Wentworth, 1838, pp. 57–104.

———. *Report on the Insects of Massachusetts, Injurious to Vegetation, Published agreeably to an order of the Legislature, by the Commissioners on the Zoological and Botanical Survey of the State*. Cambridge: Folsom, Wells, and Thruston, Printers to the University, 1841.

———. *Treatise on Some of the Insects of New England, Which Are Injurious to Vegetation*. Cambridge: John Owen, 1842.

———. *Treatise on Some of the Insects of New England Which Are Injurious to Vegetation*. 2nd ed. Boston: White & Potter, 1852.

———. *Treatise on Some of the Insects Injurious to Vegetation*. A new ed., enlarged and improved, with additions from the author's manuscripts and original notes, illustrated by engravings drawn from nature under the supervision of Professor Agassiz. Edited by Charles L. Flint. Boston: Crosby and Nichols; New York, Oliver S. Felt, 1862.

———. "Upon the Natural History of the Salt-March Caterpillar." *Massachusetts Agricultural Repository and Journal* 7, no. 4 (June 1823): 322–31.

Harvard University. *Annual Report* [1841–42]. Cambridge.

———. *Catalogue of the Library of Harvard University in Cambridge, Massachusetts . . . First Supplement*. Cambridge: C. Folsom, 1834.

———. *Catalogue of the Officers and Students* . . . (Note: 1811, 1812, 1813, 1814 are broadsides).

———. *Laws of Harvard College, For the Use of Students.* Cambridge: Hilliard and Metcalf, 1814.

———. *Statutes of the University in Cambridge Relating to the Degree of Doctor in Medicine.* Boston: Wells and Lilly, 1817.

endrickson, Walter B. "Nineteenth-Century State Geological Surveys: Early Government Support of Science." In *Science in America since 1820,* edited by Nathan Reingold, 131–45. New York: Science History Publications, 1976.

igginson, Thomas Wentworth. *Contemporaries: The Writings of Thomas Wentworth Higginson.* Boston: Houghton, Mifflin, 1900 (vol. 2: 192–218, is a reprinting, with some omissions and changes, of the biographical memoir that Higginson published in Harris, *Entomological Correspondence*).

———. "Memoir of Thaddeus William Harris." In Harris, Thaddeus William Harris. *Entomological Correspondence of Thaddeus William Harris, M.D.,* edited by Samuel H. Scudder, xi–xxxvii. Boston Society of Natural History Occasional Papers,1. Boston: Boston Society of Natural History, 1869.

ildebidle, John. *Thoreau: A Naturalist's Liberty.* Cambridge: Harvard University Press, 1983.

ill, Benjamin Thomas. *Life at Harvard a Century Ago As Illustrated by the Letters and Papers of Stephen Salisbury.* Worcester, MA: Davis Press, 1910.

istory of the Massachusetts Horticultural Society, 1829–1878. Boston: Printed for the Society, 1880.

istory of the Town of Dorchester, Massachusetts, by a Committee of the Dorchester Antiquarian and Historical Society. Boston: Ebenezer Clapp, Jr., 1859.

itchcock, Edward. *Report on the Geology, Mineralogy, Botany, and Zoology of Massachusetts, made and published by order of the government of that state: in four parts.* 2nd ed., cor. and enl. Amherst: J. S. & C. Adams, 1835.

olland, W. J. *The Moth Book: A Guide to the Moths of North America.* New York: Doubleday, Page and Co., 1903; New York: Dover Publications, 1968.

ollerbach, Anne Larsen. "Of Sangfroid and Sphinx Moths: Cruelty, Public Relations, and the Growth of Entomology in England, 1800–1840." *Osiris,* 2d ser., 11 (1996): 201–20.

olmfeld, John D. "From Amateurs to Professionals in American Science: The Controversy over the Proceedings of an 1853 Scientific Meeting." American Philosophical Society. *Proceedings* 114, no. 1 (1970): 22–36.

ooker, Joseph Dalton. *Life and Letters of Sir Joseph Dalton Hooker.* Edited by Leonard Huxley. 1918. Reprint, New York: Arno Press, 1978.

oward, L. O. "A Brief Account of the Rise and Present Condition of Official Economic Entomology." *Insect Life* 7 (1894): 58 (regarding Harris).

———. "Harris, Thaddeus William." In *Cyclopedia of American Agriculture,* 2nd ed., edited by L. H. Bailey, vol. 4: 583. New York: Macmillan, 1909–1910.

———. *A History of Applied Entomology (Somewhat Anecdotal).* Smithsonian Miscellaneous Collections vol. 84. Washington: Smithsonian Institution, 1930.

———. "Progress in Economic Entomology in the United States." *Yearbook of the United States Department of Agriculture,* 1899 (1900): 136–38 (regarding Harris).

The Insect Collection of Thaddeus W. Harris (1795–1856)." *Entomological News* 52 (December 1941): 273.

James, Mary Ann. *Elites in Conflict: The Antebellum Clash over the Dudley Observato* New Brunswick: Rutgers University Press, 1987.

Jones, William. "A New Arrangement of Papilios." Linnean Society. *Transactions* (1794): 63–69.

Judd, Richard W. *Common Lands, Common People: The Origins of Conservation in Nor ern New England.* Cambridge, MA: Harvard University Press, 1997.

Keeney, Elizabeth B. *The Botanizers: Amateur Scientists in Nineteenth-Century Ameri* Chapel Hill: University of North Carolina Press, 1992.

Kelly, Howard A. and Walter L. Burrage. *Dictionary of American Medical Biograp Lives of Eminent Physicians of the United States and Canada, from the Earliest Tim* New York: D. Appleton, 1928.

Kimball, Bruce A. *The 'True Professional Ideal' in America.* Cambridge, MA: Blackwe 1992.

Kinch, Michael Paul. "Geographical Distribution and the Origin of Life: The Dev opment of Early Nineteenth-Century British Explanations." *Journal of the Hist of Biology* 13, no. 1 (Spring 1980): 91–119

Kirby, William and William Spence. *Introduction to Entomology.* 5th ed. Londo Longman, Hurst, Rees, Orme, and Brown, 1828.

Knight, David. *Ordering the World: A History of Classifying Man.* London: Burne Books, 1981.

Kohlstedt, Sally Gregory. *The Formation of the American Scientific Community: The Am ican Association for the Advancement of Science 1848–60.* Urbana: University of I nois Press, 1976.

———. "The Nineteenth-Century Amateur Tradition: The Case of the Boston So ety of Natural History." In *Science and Its Public: The Changing Relationship,* edit by Gerald Holton and William A. Blanpied, 173–90. Dordrecht, Holland: D. R del Publishing Company, 1976.

Lankford, John. *American Astronomy: Community, Careers, and Power, 1859–1940.* Cl cago: University of Chicago Press, 1997.

Larson, Edward J. *Summer for the Gods: The Scopes Trial and America's Continuing Deb over Science and Religion.* Cambridge, MA: Harvard University Press, 1998.

Leahy, Christopher W. *An Introduction to Massachusetts Insects.* Man and Nature s ries. Lincoln: Massachusetts Audubon Society, 1983.

Lemmer, George F. "Early Agricultural Editors and Their Farm Philosophies." *A ricultural History* 31, no. 4 (October 1957): 3–22.

Lincoln, William. *History of Worcester, Massachusetts, from Its Earliest Settlement to S tember 1836.* Worcester: Moses D. Phillips and Co., 1837.

Lindroth, Carl H. "Systematics Specializes Between Fabricius and Darwin, 180 1859." In *History of Entomology,* edited by Ray F. Smith, Thomas E. Mittler, ar Carroll Smith, 119–54. Palo Alto, CA: Annual Reviews, 1973.

"List of the Writings of Thaddeus William Harris, M.D." In *Thaddeus William Ha* ris. *Entomological Correspondence of Thaddeus William Harris, M.D.,* edited by Samu H. Scudder, xxxviii–xlvii. Occasional Papers of the Boston Society of Natural H tory, 1. Boston: Boston Society of Natural History, 1869.

Lurie, Edward. *Louis Agassiz: A Life in Science.* Chicago: University of Chicago Pres 1960; Baltimore: Johns Hopkins University, 1988.

Magazine of Horticulture 9 (1843): 216–31 (review of Harris, *Treatise on Some of t Insects of New England, Which Are Injurious to Vegetation,* by "X").

Mallis, Arnold. *American Entomologists*. New Brunswick, NJ: Rutgers University Press, 1971, 25–33 (regarding Harris).

Mason, Edna Warren. *Descendants of Capt. Hugh Mason in America*. New Haven, CT: Tuttle, Morehouse and Taylor, 1937.

Massachusetts. General Court—House. House Document no. 18 (1849) (Statement of "cost of the several scientific surveys, ordered by the State, since 1830")

————. House Document no. 22 (1840) (Reports on the Reduction of Salaries and Abolishing of Commissions).

Massachusetts. *Resolves of the General Court of the Commonwealth of Massachusetts . . .* , 1838.

Matthews, J. R. "History of the Cambridge Entomological Society." *Psyche* 81, no. 1 (March 1974): 3–37.

Mawdsley, Jonathan R. "The Entomological Collection of Thomas Say." *Psyche* 100 (1993): 163–71.

McCaughey, Robert A. "The Transformation of American Academic Life 1821–1892." *Perspectives in American History* 8 (1974): 237–332.

McMurry, Sally. "Who Read the Agricultural Journals? Evidence from Chenango County, New York, 1839–1865." *Agricultural History* 63, no. 4 (Fall 1989): 1–18.

McOuat, Gordon R. "Species, Rules and Meaning: The Politics of Language and the Ends of Definitions in 19th Century Natural History." *Studies in History and Philosophy of Science* 27, no. 4 (1996): 473–519.

Meisel, Max. *A Bibliography of American Natural History: The Pioneer Century, 1769–1865*. 1924–29. Reprint, New York: Hafner Publishing, 1967.

Merrill, George P. *Contributions to a History of American State Geological and Natural History Surveys*. Smithsonian Institution, U.S. National Museum Bulletin 109. Washington: Government Printing Office, 1920.

————. *The First One Hundred Years of American Geology*. 1924. Reprint, New York: Hafner Publishing Company, 1964.

Mitchell, Lee Clark. *Witnesses to a Vanishing America: The Nineteenth-century Response*. Princeton, NJ: Princeton University Press, 1981.

Morison, Samuel Eliot. *Three Centuries of Harvard, 1636–1936*. Cambridge, MA: Harvard University Press, 1936.

Morison, Samuel Eliot and Henry Steele Commager. *The Growth of the American Republic*. New York: Oxford University Press, 1962.

Morris, John G. "Contributions toward a History of Entomology in the United States." *American Journal of Science and Arts* 51 [2nd ser., 1] (1846): 17–27.

Nash, Gerald D. "The Conflict between Pure and Applied Science in Nineteenth-Century Public Policy: The California State Geological Survey, 1860–1874." In *Science in America since 1820*, edited by Nathan Reingold, 174–85. New York: Science History Publications, 1976.

National Cyclopaedia of American Biography. New York: James T. White & Co., 1892–1984.

Neave, S. A. *The History of the Entomological Society of London, 1833–1933*. London: [Bungay, Suffolk, Printed by R. Clay & Sons], 1933.

North American Review 54 (January 1842): 73–101 (review of Harris, *Report on the Insects of Massachusetts Injurious to Vegetation;* the copy of the journal in the Widener Library, Harvard University, has a handwritten note indicating W. B. O. Peabody as author).

Novak, Barbara. *Nature and Culture: American Landscape and Painting, 1825–1875.* New York: Oxford University Press, 1980.

Nyhart, Lynn K. "Natural History and the 'New' Biology." In *Cultures of Natural History,* edited by N. Jardine, J. A. Secord, and E. C. Spary, 426–43. Cambridge: Cambridge University Press, 1996.

Ordish, George. *The Constant Pest: A Short History of Pests and Their Control.* New York: Charles Scribner's Sons, 1976.

———. *John Curtis and the Pioneering of Pest Control.* Reading, Berkshire: Osprey, 1974.

Osborn, Herbert. *Fragments of Entomological History Including Some Personal Recollections of Men and Events.* Columbus, OH: Published by the author, 1937.

Palladino, Paolo. *Entomology, Ecology and Agriculture: The Making of Scientific Careers in North America, 1885–1985.* Amsterdam: Harwood Academic Publishers, 1996.

Peabody, William B. O. *Sermons by the Late William B. O. Peabody, D.D., with a Memoir by His Brother.* 2nd ed. Boston: Benjamin H. Greene, 1849.

Porter, Charlotte M. *The Eagle's Nest: Natural History and American Ideas, 1812–1842.* University: University of Alabama Press, 1986.

"Prospectus: The Entomological Writings of Thaddeus William Harris, Edited by William Sharswood" (n.d. [ca. 1863]) (there is a copy of this item in MCZ, inscribed Dr. Hagen).

Purrington, Foster Forbes and David G. Nielsen. "Discovery of the T.W. Harris Collection at Harvard University and Designation of a Lectotype for *Podosesia Syringae* Harris." Entomological Society of Washington. *Proceedings* 89 no. 3 (1987): 548–51.

Ranz, Jim. *The Printed Book Catalogue in American Libraries: 1723–1900.* ACRL Monograph no. 26. Chicago: American Library Association, 1964.

Rehbock, Philip F. *The Philosophical Naturalists: Themes in Early Nineteenth-Century Biology.* Madison: University of Wisconsin Press, 1983.

Reingold, Nathan. "Reflections on 200 Years of Science in the United States." In *The Sciences in the American Context: New Perspectives,* edited by Nathan Reingold, 9–20. Washington, DC: Smithsonian Institution Press, 1979.

Report of the Secretary of the Class of 1863 of Harvard College, June, 1903 to June, 1913. Cambridge: University Press, John Wilson and Son, 1913.

"Report on Scientific Nomenclature." *American Journal of Science and Arts* 2nd ser., 2 (November 1846): 423–27.

Reports of the Commissioners on the Zoological Survey of the State. Massachusetts House Document No. 72. Boston: Dutton and Wentworth, 1838.

Reports of the First, Second, and Third Meetings of the Association of American Geologists and Naturalists. Boston: Gould, Kendall and Lincoln, 1843.

Roberts, Jon H. *Darwinism and the Divine in America: Protestant Intellectuals and Organic Evolution, 1859–1900.* Madison: University of Wisconsin Press, 1988.

Rosenberg, Charles. "Science in American Society: A Generation of Historical Debate." *Isis* 74 (1983): 356–67.

Rossiter, Margaret W. "The Organization of Agricultural Improvement in the United States, 1785–1865." In *The Pursuit of Knowledge in the Early American Republic: American Scientific and Learned Societies from Colonial Times to the Civil War,* edited by Alexandra Oleson and Sanborn C. Brown, 279–98. Baltimore: Johns Hopkins University Press, 1976.

Royal Society (Great Britain). *Catalogue of Scientific Papers, 1800–1900.* 19 vols. London : C. J. Clay, 1867–1925.

Russell, Charles Theodore. *Agricultural Progress in Massachusetts for the Last Half Century: An Address Delivered before the Agricultural Society of Westborough and Vicinity, September 25, 1850.* Boston: Printed by C. P. Moody, 1850.

Russell, Howard S. *A Long, Deep Furrow: Three Centuries of Farming in New England.* Hanover, NH: University Press of New England, 1976.

Sammarco, Anthony Mitchell. *Dorchester.* Images of America. Dover, NH: Arcadia, 1995.

———. *Dorchester.* Vol. 2. Images of America. Dover, N.H.: Arcadia, 2000.

Sawyer, Richard C. *To Make a Spotless Orange: Biological Control in California.* Henry A. Wallace Series on Agricultural History and Rural Life. Ames: Iowa State University Press, 1996.

Say, Thomas. *The Complete Writings of Thomas Say on the Entomology of North America.* Edited by John L. LeConte. New York: Bailliere Brothers; London, H. Bailliere, 1859.

———. "Descriptions of New North American Insects, and Observations on Some Already Described." American Philosophical Society. *Transaction,* n.s., 6 (1839): 155–90.

Schlesinger, Jr., Arthur M., gen. ed. *The Almanac of American History.* New York: Putnam, 1983; New York: Bramhall House, 1986.

Scudder, Samuel H. *Butterflies of the Eastern United States and Canada.* Cambridge [MA]: Published by the Author, 1889), vol. 1:656–58 (regarding Harris).

Scudder, Samuel H. [editor]. "Some Old Correspondence between Harris, Say, and Pickering." *Psyche* 6 (1891–93): 57–60, 121–24, 137–41, 169–72, 185–87, 357–58.

Shipton, Clifford K. *Biographical Sketches of Those Who Attended Harvard College in the Classes 1761–1771.* Sibley's Harvard Graduates 17. Boston: Massachusetts Historical Society, 1975, pp. 611–13 (entry on Reverend Zedekiah Sanger).

Sibley, John Langdon. Collectanea Biographica Harvardiana [scrapbook of newspaper obituaries and related death notices]. John Langdon Sibley Papers, Harvard University Archives, HUG 1791.14, vol. 1:225 (obituaries of Harris).

Slotten, Hugh Richard. *Patronage, Practice, and the Culture of American Science: Alexander Dallas Bache and the U.S. Coast Survey.* Cambridge: Cambridge University Press, 1994.

Smallwood William Martin, with Mabel Sarah Coon Smallwood. *Natural History and the American Mind.* New York: Columbia University Press, 1941.

Smith, James E. and John Abbot. *The Natural History of the Rarer Lepidopterous Insects of Georgia . . . Collected from the Observations of Mr. John Abbot.* London: Printed by T. Bensley, for J. Edwards, Cadell and Davies; and J. White, 1797.

Smith, John B. "A Monograph of the Sphingidae of America North of Mexico." American Entomological Society. *Transactions* 15 (April 1888): 49–242.

Snow, Caleb H. *A History of Boston . . . with Some Account of the Environs.* Boston: Abel Bowen, 1825.

Sorensen, W. Conner. *Brethren of the Net: American Entomology, 1840–1880.* Tuscaloosa: University of Alabama Press, 1995.

Stanton, William. *The Great United States Exploring Expedition of 1838–1842.* Berkeley and Los Angeles: University of California Press, 1975.

Sterling, Keir B., Richard P. Harmond, George A. Cevasco, and Lorne F. Ham monds, eds. *Biographical Dictionary of American and Canadian Naturalists and Env ronmentalists.* Westport, CT: Greenwood Press, 1997.

Stone, Bruce Winchester. "The Role of the Learned Societies in the Growth of Sci entific Boston 1780–1848." Ph.D. diss., Boston University, 1974.

Strecker, Herman. *Butterflies and Moths of North America.* Reading, PA: Press of B. F Owen, 1878, 238–40 (bibliography of works by Harris).

Stroud, Patricia Tyson. *Thomas Say: New World Naturalist.* Philadelphia: University o Pennsylvania Press, 1992.

"Thaddeus William Harris, M.D. [note accompanying portrait]." *Entomologica News* 7, no. 1 (January 1896): 1.

Thoreau, Henry David. "Natural History of Massachusetts." *The Dial: A Magazine fo Literature, Philosophy, and Religion* 3 (July 1842): 19–40.

―――. *The Writings of Henry David Thoreau,* vol. 7–20, *Journal.* Edited by Bradforc Torrey. Boston: Houghton Mifflin; Cambridge: Riverside Press, 1906.

Thornton, Tamara Plakins. *Cultivating Gentlemen: The Meaning of Country Life amon the Boston Elite 1785–1860.* New Haven: Yale University Press, 1989.

Tichi, Cecelia. *New World, New Earth: Environmental Reform in American Literature fror the Puritans through Whitman.* New Haven: Yale University Press,1979.

Tuxen, S. I. "Entomology Systematizes and Describes, 1700–1815." In *History of En tomology,* edited by Ray F. Smith, Thomas E. Mittler, and Carroll Smith, 95–118 Palo Alto, CA: Annual Reviews, 1973.

Vaille, F. O. and H. A. Clark. *The Harvard Book: A Series of Historical, Biographical and Descriptive Sketches.* 2 vols. Cambridge: Welch, Bigelow and Company, 1875 available as online resource from the Harvard University Archives website a http://hul.harvard.edu/huarc/

Van Anglen, Kevin P. "True Pulpit Power: 'Natural History of Massachusetts' anc the Problem of Cultural Authority." In *Studies in the American Renaissance 1990* edited by Joel Meyerson, 119–47. Charlottesville: University Press of Virginia 1990.

Veysey, Laurence. "Higher Education as a Profession: Changes and Continuity." Ir *The Professions in American History,* edited by Nathan O. Hatch, 15–32. Notre Dame, IN: University of Notre Dame Press, 1988.

Wade, J. S. "The Friendship of Two Old-Time Naturalists." *Scientific Monthly* 22 (Au gust 1926): 152–60.

Walton, Clarence E. *The Three-Hundredth Anniversary of the Harvard College Library* Cambridge, MA: Harvard College Library, 1939.

Weiss, Harry B. *The Pioneer Century of American Entomology.* New Brunswick, NJ: Pub lished by the author [Mimeographed], 1936.

Weiss, Harry B., and Grace M. Ziegler. *Thomas Say: Early American Naturalist.* Spring field, IL: Charles C. Thomas Publisher, 1931.

Welch, Margaret. *The Book of Nature: Natural History in the United States, 1825–1875* Boston : Northeastern University Press, 1998.

Wheatland, David P. *The Apparatus of Science at Harvard 1765–1800.* [Cambridge] Harvard University Collection of Historical Scientific Instruments, 1968.

Wheeler, Jr., A. G. "The Tarnished Plant Bug: Cause of Potato Rot?—An Episod in Mid-Nineteenth-Century Entomology and Plant Pathology." *Journal of the His tory of Biology* 14, no. 2 (Fall 1981): 317–38.

Whorton, James. *Before Silent Spring: Pesticides and Public Health in Pre-DDT America.* Princeton, NJ: Princeton University Press, 1974.

Winsor, Mary P. *Starfish, Jellyfish, and the Order of Life: Issues in Nineteenth-Century Science.* New Haven: Yale University Press, 1976.

Zochert, Donald. "Science and the Common Man in Antebellum America." In *Science in America since 1820,* edited by Nathan Reingold, 7–32. New York: Science History Publications, 1976.

Index

Page numbers in italics refer to the illustrations.